Charles S. Peirce
on the
Logic
of
NUMBER

PAUL SHIELDS

Docent
Press

DOCENT PRESS
Boston, Massachusetts, USA
www.docentpress.com

Docent Press publishes monographs and translations in the history of mathematics for thoughtful reading by professionals, amateurs and the public.

Cover design by Brenda Riddell, Graphic Details, Portsmouth, New Hampshire.

Preface

This book was originally a doctoral dissertation, written under the direction of Vincent G. Potter, S. J., at Fordham University in 1981. It has been edited to improve its clarity and readability, and the improvised mathematical notation of the typescript has been changed into standard mathematical notation. This preface is new, and there is a new appendix containing a few letters from Max Fisch which might be of interest to Peirce scholars. Otherwise the content is essentially unchanged. It contains a proof of the equivalence of the Peirce and Dedekind axioms for the natural numbers – a proof that has hitherto been known mainly by report.[1] Awareness that Peirce's axiom system exists has been gradually increasing – the current Wikipedia entry on the Peano Axioms, for instance, cites it prominently – but the conventional wisdom among mathematicians is still that the first satisfactory axiom systems were those of Dedekind and Peano.[2]

My work was set against the background of an earlier generation of Peirce scholarship: Max Fisch, Murray G. Murphey, Thomas Goudge, Carolyn Eisele, Vincent Potter, and others – scholars from whom I learned much. Murphey, in particular, impressed upon me the relevance of Peirce's mathematical thought to his philosophy as a whole, and introduced me to the question of Georg Cantor's influence upon Peirce. But over the last thirty years, with the gradual publication of the chronological *Writings of Charles S. Peirce* by the Peirce Edition Project,[3] a new generation of scholarship has emerged with a better understanding of the significance of Peirce's mathematical thought. I am less acquainted with this new scholarship, but its quality and scope are evident in the recent *New Essays on Peirce's Mathematical Philosophy*, edited by Matthew Moore.[4]

My dissertation followed Fraenkel and Bar-Hillel's *Foundations of Set Theory* in distinguishing "Hilbert arithmetics" from "Peano arithmetics".[5] This usage is rare today, so I have taken the liberty of replacing these designations with the more common: "first-order Peano arithmetics" and "second-order Peano arithmetics." First-order arithmetics can be formulated using only first-order predicate logic; second-order arithmetics require a logic which permits quantification over predicates. To the extent that this distinction can be applied ahistorically, the axiom systems of Peirce, Dedekind, and Peano are evidently

---

[1][**115**]

[2]Wikipedia, "Peano Axioms," http://en.wikipedia.org/wiki/Peano_axioms (accessed July 30, 2012).

[3][**87**]

[4][**68**]

[5]See [**31**, p. 292 ff.].

second-order arithmetics, since their mathematical induction axioms are construed to hold for *all* classes or properties of numbers. In an important sense, however, the distinction itself has been a twentieth century creation. Like axiomatic set theory, it arose in response to the paradoxes. My discussion cites W. V. O. Quine's contention that second-order logic is merely "set theory in sheep's clothing." But Quine's view was challenged in the 1980's and 1990's by the logician George Boolos and others, and I would refer the interested reader to these new developments.[6]

Two excellent recent essays on Peirce bear, in different ways, upon the present work. The first is Jérôme Havenel's "Peirce's Clarifications of Continuity," which appeared in 2008.[7] It revises Potter's and my work on Peirce's definitions of continuity,[8] providing more extensive context and a more precise chronology. Havenel's essay should be a valuable resource for future scholars. The second is Shannon Dea's "'Merely a veil over the living thought': Mathematics and Logic in Peirce's Forgotten Spinoza Review," from 2006.[9] This elegant essay presents Peirce's understanding of mathematics as being "from the inside out" – a description that seems to me consistent with the sketch given below (in section 3.3). In fact, if there were anything that I would approach differently today, it would probably be to emphasize, even more than I do, the critical importance of Benjamin Peirce to Charles' understanding of mathematics.

I owe a tremendous debt to Father Potter for his support during the writing of the dissertation. He was a wonderful scholar – and was my mentor and my friend. I was fortunate to have perceptive readers in Elizabeth Kraus and Gerrit Smith. They were also outstanding teachers, as were others with whom I studied at this time, including John Bacon, Vincent Cooke, and John Smith. I can still hear their voices in my writing. Max Fisch was characteristically generous with his time and with his extensive knowledge of Peirce. Philosophy often occurs in conversation among friends, and for this I am grateful to Andrew Chrucky, Frank Purcell, Dominic and Mamie Balestra, Donna Orange, Janice Shields, and many others. The dissertation was supported by a fellowship, during 1979–80, from the Deutscher Akademisher Austauschdienst.

---

[6]See [**13**], [**60**], and [**120**].
[7][**42**]
[8][**91**]
[9][**22**]

I would like to thank Philip Shields and Shelley Hope for their support and encouragement during these revisions. I also want to thank Scott Guthery, my very patient editor.

Paul Shields
October 2012

Contents

Introduction

In 1881 the American philosopher Charles S. Peirce published a remarkable paper in *The American Journal of Mathematics* called "On the Logic of Number." Peirce's paper marked a watershed in nineteenth century mathematics, providing the first successful axiom system for the natural numbers. Since scholarship has traditionally attributed priority in this regard to the axiom systems of Richard Dedekind, in 1888, and Giuseppe Peano, in 1889, we will show that Peirce's 1881 axiom system is actually equivalent to these better known systems. The significance of this, in part, will be to give credit where credit is due. But it will also bring into clearer focus the growth of mathematical abstraction in the nineteenth century. To this end, we will discuss striking technical features in Peirce's approach. It is not generally known, for instance, that his 1881 paper provided the first abstract formulation of the notions of partial and total linear order, that it introduced recursive definitions for arithmetical operations, nor that it proposed the first general definition of cardinal numbers in terms of ordinals.

Peirce was probably America's greatest philosopher, and his interest in the foundations of mathematics was closely tied to his main philosophical concerns. Some of his most characteristic philosophical positions – his synechism and his phenomenological categories – bear the direct imprint of his research into the theory of sets and transfinite numbers. Peirce's 1881 paper, in particular, is important for understanding his view of the nature of mathematics and its relation to deductive logic. It was published concurrently with his father's famous definition of mathematics as *the science which draws necessary conclusions*. This historical juxtaposition is significant since it indicates how Peirce would have elaborated the philosophical implications of his 1881 axiom system.

In the course of tracing out the implications of Pcirce's 1881 paper, we address the problem of locating his mature philosophy of mathematics vis-à-vis the traditional triad of logicism, formalism, and intuitionism. Although we show that Peirce's view had similarities to and differences from all three, his understanding of mathematics was essentially *sui generis*. Perhaps the most characteristic aspect of Peirce's approach is that he did not conceive mathematics to require any sort of epistemological foundation, whether in logic, intuition, or by means of constructive completeness proofs. This is why Peirce, in his scheme of categories, characterized mathematics as a First. "There is no more satisfactory way of assuring ourselves of anything," Peirce said, "than the mathematical way of assuring ourselves of mathematical theorems."

Our presentation is divided into three chapters. Chapter I is basically an introduction, both historical and conceptual, to the axiomatization of arithmetic. It begins with three historical sections which extrapolate the essential ideas from, respectively, "On the Logic of Number," Dedekind's *Was sind und was sollen die Zahlen?*, and Peano's *Arithmetices Principia Nova Methodo Exposita*. Special attention is paid in these sections to issues of terminology and notation. The next sections – on transitive relations and functions, simple order and chains, mathematical induction and arithmetical operations – are more conceptually oriented and focus upon the contrasting approaches of Peirce and Dedekind to the requirements of axiomatization. The first chapter concludes with three sections – on categoricity, cardinality, and the existence of a model – which are concerned more with the adequacy than the mechanics of axiomatization.

Chapter II is the core of the book, giving a formal proof of the equivalence of the axiom systems of Peirce and Dedekind. Portions of the proof reformulate theorems given in *Was sind und was sollen die Zahlen?*. The significance of the remainder is clearly historical rather than mathematical, establishing the priority of Peirce with regard to one of the most fundamental accomplishments in modern foundations.

Chapter III contains an overall reappraisal of Peirce's 1881 paper, treating at greater length some of the themes already mentioned in this introduction. It is divided into three sections. The first examines the mathematical context of Peirce's paper, and the second provides a reevaluation of its importance for the the foundations of mathematics. The third section constitutes a distinct essay which attempts to describe the outlines of Peirce's philosophy of mathematics. It takes into consideration his 1881 axiom system, but also draws upon his broader thought and his writing from other periods in his life.

For many years the most valuable primary source on Peirce has been the *Collected Papers of Charles Sanders Peirce* [84], [85]. We will follow the general practice of Peirce scholarship in noting references to *The Collected Papers* by volume and paragraph number in the body of the text. For example, (5.402) will refer to volume 5, paragraph 402. An additional source, often used by Peirce scholars, is the collection of manuscripts and letters housed in Houghton

Library at Harvard University. These were microfilmed in 1963–64 by the Harvard University Microreproduction Service. With the addition, in 1967, of Richard S. Robin's invaluable *Annotated Catalogue of the Papers of Charles S. Peirce* [**102**], general access to these manuscripts and letters became possible. References to them will follow the numbering given in Robin's *Catalogue,* or his "The Peirce Papers: A Supplementary Catalogue" [**103**]. For example, (Ms 28) will refer to the manuscript with the Robin number of 28.[10]

---

[10]Future Peirce scholarship will also have available the projected thirty volume *Writings of Charles S. Peirce: A Chronological Edition,* currently being published by the Peirce Edition Project [**87**].

CHAPTER 1

The Axiomatization of Arithmetic

This chapter will provide an informal introduction to the first axiom systems for elementary arithmetic. We will begin by describing the axiomatizations of Peirce, Dedekind, and Peano, and we will then discuss some of the basic concepts and techniques of these systems, focusing upon a comparison of the approaches advocated by Peirce and Dedekind. The related problems of categoricity, cardinality, and the existence of an appropriate model will also be introduced.

1.1. "On the Logic of Number"

In 1881, while teaching at Johns Hopkins, Peirce published a paper in *The American Journal of Mathematics* entitled "On the Logic of Number."[1] In its attempt to construct an axiomatic basis for arithmetic, Peirce's 1881 paper was prefigured by an earlier 1867 paper called "Upon the Logic of Mathematics."[2] This earlier paper used a rather primitive Boolean calculus and Peirce later dismissed it as "by far the worst I ever published." (4.333) In 1881 Peirce brought to the task of axiomatization a much more powerful tool – the logic of relations – and wrote a paper with which he remained satisfied his entire life.[3]

The first paragraph of "On the Logic of Number" sets forth Peirce's program:

---

[1] [**76**, pp. 85–95], reprinted at 3.252-288. This paper antedated the main period (1890-1910) of Peirce's preoccupation with set theoretical topics.

[2] Proceedings of the *American Academy of Arts and Sciences* 7 (1867): 402-412; reprinted at 3.20-44

[3] The extent to which Peirce's 1881 paper remained central to his later thought will be shown in Chapter 3. The importance of Peirce's development of the logic of relations for subsequent foundations work is emphasized by Bertrand Russell in *Principles of Mathematics* (London, 1903; paperback ed. New York. W. W. Norton & Co., n.d.), pp. 23–24. On Peirce's apparently independent discovery of this logic, see Emily Michael, "Peirce's Early Study of the Logic of Relations, 1865-1867," *Transactions of the Charles S. Peirce Society* 10 (1974). 63-75.

Nobody can doubt the elementary propositions concerning number: those that are not at first sight manifestly true are rendered so by the usual demonstrations. But although we see they *are* true, we do not so easily see precisely *why* they are true; so that a renowned English logician has entertained a doubt as to whether they were true in all parts of the universe. The object of this paper is to show that they are strictly syllogistic consequences from a few primary propositions. The question of the logical origin of these latter, which I here regard as definitions, would require a separate discussion. (3.252)

Peirce intends to show that elementary arithmetical propositions are "strictly syllogistic consequences from a few primary propositions." These primary propositions, or axioms, are introduced in five successive paragraphs:

Quantity. Let r be any relative term, so that one thing may be said to be r of another, and the latter r'd by the former. If in a certain system of objects whatever is r of an r of anything is itself r of that thing, then r is said to be a transitive relative in that system ... In a system in which r is transitive, let the q's of anything include that thing itself, and also every r of it which is not r'd by it. Then q may be called a fundamental relative of quantity; its properties being, first, that it is transitive; second, that everything in the system is q of itself, and third, that nothing is both q of and q'd by anything except itself. The objects of a system having a fundamental relation of quantity are called quantities, and the system is called a system of quantity. (3.253)

Simple Quantity. A system in which of every two quantities one is a q of the other is termed simple. (3.254)

Discrete Simple Quantity. A system of simple quantity is either continuous, discrete, or mixed ... A discrete system is one in which every quantity greater than another is next greater than some quantity (that is, greater than without being greater than something greater than.) (3.256)

Semi-limited Discrete Simple Quantity. A simple system of discrete quantity is either limited, semi-limited, or unlimited. A limited system is one which has an absolute maximum and an absolute minimum quantity; a semi-limited system has one (generally considered a minimum) without the other; an unlimited has neither. (3.257)

Infinite Semi-limited Discrete Simple Quantity. A simple, discrete, system, unlimited in the direction of increase ... is in that direction either infinite or super-infinite... An infinite class is one in which if it is true that every quantity next greater than a quantity of a given class itself belongs to that class, then it is true that every quantity greater than a quantity of that class belongs to that class. (3.258)

These five definitions form the axiomatic basis for what Peirce calls "ordinary number": a system of infinite semi-limited discrete simple quantity (3.260)

Since Peirce's terminology does not necessarily accord with modern usage, it will be useful to give a preliminary explanation of these axioms:

1) Peirce's first axiom describes a relation – think of it as the relation "greater than or equal to" – which is transitive, antisymmetric, and reflexive in a given system of objects. Such a system Peirce calls a system of quantity. It is what we would today call a partially ordered set.

2) Peirce's second axiom adds the requirement that a system of quantity be connected, and he calls the result a system of simple quantity. A set that is both partially ordered and connected by a given relation is commonly referred to today as being simply, totally, or linearly ordered.

3) Peirce's third axiom describes a system in which every element except the minimum has an immediate predecessor. His use of the terms "discrete" and "continuous" in this context is somewhat unfortunate.[4] But he clearly intends to require that ordinary numbers be closed with respect to immediate predecessors.

4) Peirce's fourth axiom is straightforward. It requires that the system have a minimum element but no maximum element.

5) Peirce's final axiom has nothing to do with being infinite as opposed to finite, but is intended to isolate from among those systems which are infinite those which are not, as Peirce puts it, "super-infinite." It states the principle of mathematical induction, the form of inference that Peirce often called "Fermatian inference." Because of how he

---

[4]Peirce's understanding of continuity, in 1881, is still rather rudimentary. See my paper with Vincent G. Potter, [**91**]. Also see [**42**] which provides more extensive context and a more precise chronology for the changes in Peirce's thinking.

formulates the principle, we will refer to it as "mathematical induction starting with k." It is important not to confuse this use of the
term "infinite" with the discussion of the cardinal definition of finitude
which occurs later in Peirce's paper. It is this later cardinal definition
over which the dispute with Dedekind arose.[5]

Each of the non-primitive terms introduced, e.g., "greater than," "next greater
than," "minimum," "maximum," can be defined in terms of Peirce's fundamental relation of quantity, q.[6] Peirce's entire axiom system can be compactly
expressed in contemporary terms:

> Let N be a set, R a relation in N, and assume the normal
> definitions of minimum and maximum:
> 1. N is partially ordered by R
> 2. N is connected by R
> 3. N (except for 1) is closed with respect to predecessors
> 4. 1 is the minimum element in N; N has no maximum
> 5. Mathematical induction starting with k holds for N.

The central section of Peirce's paper describes the arithmetic of the natural
numbers. Peirce calls the minimum element "1", and then establishes recursive
definitions for addition and multiplication:

> By $x + y$ is meant, in case $x = 1$, the number next greater than
> y; and in other cases, the number next greater than $x' + y$,
> where x' is the number next smaller than x.
>   By $x \times y$ is meant, in case $x = 1$, the number y, and in
> other cases $y + (x' \times y)$, where x' is the number next smaller
> than x.

---

[5]These are apparently confused by the editors of the Collected Papers, since they mis-
leadingly cross reference the 3.258 definition, above, with a passage at 3.564 which clearly
refers only to the later cardinal definition at 3.288. Concerning the dispute with Dedekind
see below, pp. 56ff.

[6]The definitions based on q are as follows:

1. x is greater than y iff x is q of y and y is not q of x.
2. x is next greater than y iff
   i. x is q of y
   ii. y is not q of x, and
   iii. $\forall z$, if z is q of y and y is not q of z, then z is q of x.
3. x is a minimum iff x is q only of itself.
4. x is a maximum iff x is q'd only by itself.

It may be remarked that the symbols + and × are triple relatives, their two correlates being placed one before and the other after the symbols themselves. (3.262-264)

Again, it is easy to see how the relation "next smaller than" can be defined in terms of q. Peirce goes on to prove associative, commutative, and distributive laws for these operations. He also proves a few theorems for an unlimited system, i.e., the system of integers.

Finally, two passages near the end of "On the Logic of Number" break entirely new ground. The first introduces the notion of the "cardinality" of a set – roughly, the "amount" of elements it contains – by describing the process of counting

Let such a relative term c that whatever is a c of anything is the only c of that thing, and is a c of that thing only, be called a relative of simple correspondence. ...

If every object, s, of a class is in any such relation being c'd by a number of a semi-infinite discrete simple system, and if further every number smaller than a number c of an s is itself c of an s, then the numbers c of the s's are said to count them, and the system of correspondence is called a count ... If in any count there is a maximum counting number the count is said to be finite, and that number is called the number of the count. (3.280)

A "relative of simple correspondence" establishes what would today be called a "one-to-one correspondence." Peirce is proposing that the cardinality of a set be determined by placing it in one-to-one correspondence with an initial segment of natural numbers.[7] A set will be finite, by this proposal, just in case the corresponding initial segment has a maximum.

The second passage provides a contrasting, purely cardinal definition of what it means to be a finite set. It is based upon De Morgan's "syllogism" of transposed quantity.[8]

---

[7]This procedure, which probably originated with Peirce, underlies the modern method of defining cardinals as initial ordinals. See [**41**, pp. 99–100] below, pp. 47–56. The axiom of choice is required to prove that every set has such a cardinal number.

[8]This "syllogism" is actually a paralogism in the treatment of De Morgan, since the premiss of finitude is omitted. See Peirce's discussion at 3.402.

If every S is P, and if the P's are a finite lot counting up to a number as small as the number of S's, then every P is an S. For if, in counting the P's, we begin with the S's (which are a part of them), and having counted all the S's arrive at the number n, there will remain over no P's not S's. For if there were any, the number of P's would count up to mere than n. From this we deduce the validity of the following mode of inference:

> Every Texan kills a Texan,
> Nobody is killed by but one person,
> Hence, every Texan is killed by a Texan,

supposing Texans to be a finite lot . . . . This mode of reasoning is frequent in the theory of numbers. (2.288)

A finite set, as implied by this second proposal, would be one for which the above reasoning – the syllogism of transposed quantity – is valid. This passage marks the first appearance in Peirce's published work of this famous definition, which is stated explicitly four years later in Pierce's paper "On the Algebra of Logic: A Contribution to the Philosophy of Notation."[9]

Notice that both of the above passages imply a corresponding definition of the infinite, i.e., an infinite set is one that is not finite. It is important to emphasize, however, that these definitions are not equivalent. The first depends upon Peirce's previously defined system of "ordinary numbers", while the second is purely cardinal in that it can be stated without any reference to such a system. The properties defined are commonly referred to in modern set theory as those of being, respectively, "finite" and "Dedekind finite." In the cited passage, Peirce derives the latter from the former. It is known today that to establish their equivalence by making the reverse derivation requires the axiom of choice.[10]

---

[9][**80**, 180–202], reprinted at 3.359–403. In particular see 3.402.

[10]The first complete survey of various definitions of finitude in relation to the axiom of choice was [**118**]. On the two definitions above, see [**117**, pp. 98, 99, 107, 240, 241]; or the informal treatment in [**112**, ch. 12]

1.2. Was sind und was sollen die Zahlen?

Richard Dedekind's essay, published in 1888, is generally considered one of the pioneering works in the foundations of mathematics.[11] We will summarize the central ideas of this work and try to remove obstacles to a contemporary understanding of it.

In his first preface, Dedekind describes *Was sind und was sollen die Zahlen?* as "an attempt to establish the science of numbers upon a uniform foundation."[12] And this "uniform foundation" is clearly meant to include an axiom system for elementary arithmetic. But in contrast to Peirce, whose reliance upon the logic of relations was only casually augmented by references to "systems" and "classes," Dedekind adopts set-theoretical methods throughout his work – initiating an important new approach to the foundations of mathematics.[13] The first section of his essay (§1–§20) introduces the general theory of sets [*Systeme*]. Set equality is defined extensionally, i.e., two sets are equal. when they contain the same elements, and the usual definitions of a subset [*Teil*] and proper subset [*echter Teil*] are given. These are followed by definitions of the union [*zusammengesetzten System*] and intersection [*Gemeinheit*] of a family of sets.[14]

There are several inaccuracies in Dedekind's treatment that ought to be mentioned immediately. To begin with, Dedekind does not include the empty set in his foundation, even though he recognizes that it might be useful for other investigations.[15] Secondly, he does not always distinguish between membership in a set and being a subset of that set, using the same symbol for both relations. For example, Dedekind seems to define the concept of a "unit set"

---

[11]See [**24**]. The theorems and definitions in Dedekind's book are numbered consecutively, and we will cite these numbers parenthetically in the text. E.g., Dedekind's twenty second theorem will be cited in the text as §22.

[12]See [**24**, p. 34].

[13]Dedekind's more or less intuitive set theory was utilized in the first axiomatization, by Ernst Zermelo, two decades later, Ernst Zermelo[**127**].

[14]Exact definitions of these concepts are given in Chapter 2. Dedekind's symbol for set inclusion, "3" will be replaced by the usual symbol, $\subseteq$ and proper subset will be indicated by the symbol, $\subset$. Similarly, the notation Dedekind uses for unions and intersections will be replaced by the modern symbols, $\cup$ and $\cap$.

[15]What Dedekind says is: "Dagegen wollen wir das leere System ...aus gewissen Gründen hier ganz ausschliessen, obwohl es für andere Untersuchungen bequem sein kann, ein solches zu erdichten" (§2) A critical reconstruction of just what these *gewissen Gründen* are can be found in [**34**, pp. 149–150]. We will require at least the uniqueness of such a set for our formalization in Chapter 2.

> ...it is advantageous to include also the special case where a
> set S consists of a *single* (one and only one) element a, i.e.,
> the thing a is an element of S but everything different from a
> is not an element of S. (§2)

but he then goes on to say that "every element s of a set S ... can itself be
regarded as a set," concluding that we might as well describe, notationally, such
elements as being subsets of S (§3). As a result, Dedekind conflates, throughout
his work, elements with their unit sets.[16] And thirdly, Dedekind's definitions of
union and intersection, as well as many of the proofs involving these concepts,
are faulted by reliance on ellipses.[17] These difficulties are technical in nature and
do not affect the underlying sense of Dedekind's axiomatization. Nonetheless,
they will require appropriate modifications in our subsequent presentation.

Dedekind's second section (§21–§25) introduces the notion of a function
[*Abbildung*]. With respect to a function f, the image [*Bild*], M', of a set M is
defined to consist of those elements $f(x)$ such that x is in M.[18] The concept
of a one-to-one function [*ähnliche Abbildung*] is presented in Dedekind's third
section (§26–§35).

The fourth section (§36–§63) is concerned with functions which map sets
into themselves. A set, M, is called a chain [*Kette*] with respect to a function
when it contains its own image, i.e. when $M' \subseteq M$.[19] This concept, of a
set whose image is a subset of itself, gained currency through its philosophical
adaptation by Josiah Royce in his "Supplementary Essay" to *The World and
the Individual*. Royce gave the graphic example of a map of England drawn
upon the surface of England: if accurate enough, such a map would include
a map of itself, the map of the map again including a map of itself, etc.[20] It
should be noted, though, that Royce's conception of a *Kette* differs from that

---

[16]*Ibid.*, p. 149. This conflation affects the statement, below, of Dedekind's second axiom.

[17]*Ibid.*, p. 139.

[18]Exact definitions of these concepts will be given in Chapter 2. Beman translates
Abbildung and *Bild* as "transformation" and "transform." [**24**, p. 50]. Bertrand Helm, in
[**45**, p. 232] claims that a different translation as, respectively, "representation" and "image,"
given in [**107**, pp. 511ff], is "logical" instead of "mathematical," and responsible for what
Helm takes to be Royce's misunderstanding of Dedekind. Royce's translation, however, is also
adopted by Bertrand Russell [**109**, pp. 245–251] and by Peirce at 3.609. Our own translation
most accurately reflects modern usage – see [**117**, pp. 65, 86]. Note that Dedekind speaks of
the "images" of elements as well as of sets (§21), a practice we shall avoid.

[19](§37). Note that this differs from the common modern usage of the term "chain," see
the discussion below, pp. 24–27.

[20]See [**102**, 526-538]. Royce uses this notion-to characterize the abstract structure of an
ideal self. It is also adopted by Peirce, as a metaphor for self-consciousness, at 5.71.

of Dedekind in being confined to those with respect to functions which are one-to-one.[21]

A useful theorem on chains (§43) is that the intersection of a family of chains must itself be a chain.[22] This theorem helps elucidate the most important definition (§44) of the section, let A be a set; the chain of set A [die *Kette* des Systems A] is denoted by $A_\circ$, and is defined as the intersection of all those chains which contain A. Alternatively, $A_\circ$ is implicitly defined by the following three theorems; (§45) $A \subseteq A_\circ$, (§46) $A_\circ' \subseteq A_\circ$, and (§47) $A \subseteq K \ \& \ K' \subseteq K \Rightarrow A_\circ \subseteq K$, assuming only that the family of chains including A is not empty.[23]

As an example of the chain of a set, consider the ordinal numbers and let f be the function which maps each ordinal onto its immediate successor. These numbers clearly constitute a chain with respect to f, since the immediate successor of any ordinal is also an ordinal. Now let A be any set of natural numbers of which the number, n, is the least. Then $A_\circ$ will be the set of all natural numbers greater than or equal to n, i.e., $A_\circ$ will be the smallest chain with respect to f containing A.[24]

The importance of this definition is that it allows Dedekind to derive a generalized form of the principle of mathematical induction, which he calls "the theorem of complete induction" (§59). This theorem asserts that, given a function f and two sets, A and M, if A is a subset of M, and if the image of the intersection of $A_\circ$ and M is also a subset of M, then $A_\circ$ is itself a subset of M. The proof of this powerful theorem on the basis of the few definitions preceding it is one of the more elegant features of Dedekind's essay.[25]

The fifth section (§64–§70) begins with Dedekind's cardinal definition of what it means to be an infinite set. A set is called infinite if it can be put into one-to-one correspondence with a proper subset of itself, and finite otherwise (§64). This is the definition later shown by Ernst Schröder to be equivalent to Peirce's definition using the syllogism of transposed quantity.[26] Also included in this section is Dedekind's attempt to prove that such infinite sets exist.[27]

---

[21]Compare [**107**, 502–509] with Dedekind §37.

[22]See Chapter 2, Theorem T5.

[23]An excellent contemporary exposition of this concept is that given in [**11**, pp. 30–31].

[24]This example is taken from Bertrand Russell's *Principles of Mathematics,* p. 246. Although the function it employs is one-to-one, again, Dedekind's definition of $A_\circ$ does not specify that the function must be one-to-one.

[25]See Chapter 2, Theorem T11.

[26]See [**113**, pp. 303–315]

[27]See below, pp. 56ff.

The sixth section (§71–§80) gives Dedekind's axioms for the natural numbers.

A set, N, is said to be simply infinite (§71) when there is a function, f, such that:

    i. N is a chain

    ii. N is the chain of the unit set $\{1\}$

    iii. 1 is not an element of N', and

    iv. f is one-to-one.[28]

Since different sets can satisfy these conditions, Dedekind suggests that the special characteristics of their elements be ignored and that only the relationship between elements, as established by the function f, be considered. Taken in this sense, the elements of a simply infinite system are called natural numbers [*natürliche Zahlen*] or ordinal numbers [*Ordinalzahlen*] (§73). Dedekind's famous observation follows: "With reference to this freeing of the elements from every other content (abstraction) we are justified in calling numbers a free creation of the human mind."[29]

The remainder of Dedekind's essay concerns various topics which will be examined in greater detail later in this chapter. The seventh and eighth sections (§81–§123)) define the transitive relations of "greater" and "lesser" among natural numbers, and the initial segment, Z_n, corresponding to each natural number, n. The ninth section (§124–§131) introduces recursive definitions. The tenth section, (§132–§134) proves that any two simply infinite systems are isomorphic. Sections eleven through thirteen (§135–§158) define the arithmetical operations – addition, multiplication, and exponentiation – and derive the expected laws. Finally, Dedekind's fourteenth section (§159–§172) defines the cardinality [*Anzahl, Grad*] of a finite set as the number, n, such that the set can be put into one-to-one correspondence with the initial segment, Z_n (§161). Unlike Peirce, Dedekind attempts to prove the full equivalence of his earlier cardinal definition of the infinite and the ordinary definition according to which a set is infinite when it cannot be put into one-to-one correspondence with any initial segment, Z_n (§159). His proof, however, implicitly assumes the axiom of choice.[30]

---

[28]Dedekind states his second axiom as $N = 1_o$, but only defines the chains of *sets*. Our formalization in Chapter 2 will modify this axiom.

[29]"In Rücksicht auf diese Befreiung der Elemente von jedem anderen Inhalt (Abstraktion) kann man die Zahlen mit Recht eine freie Schöfung des menschlichen Geistes nennen."

[30]A third (cardinal) definition of the infinite, given in Dedekind's second preface, *Essays*, p. 41, also requires the axiom of choice to show its equivalence to §64.

1.3. Arithmetices principia nova methodo exposita

Giuseppe Peano's little booklet was published in 1889, a year after the publication of *Was sind und sollen die Zahlen?*.[31] Like the programs of Peirce and Dedekind, which employed, respectively, the logic of relations and new set-theoretical methods, Peano's *Arithmetices principia* shows an increased technical sophistication. He begins his booklet by saying:

> Questions pertaining to the foundations of mathematics, although treated by many these days, still lack a satisfactory solution. The difficulty arises principally from the ambiguity of ordinary language. For this reason it is of the greatest concern to consider attentively the words we use. I resolved to do this, and am presenting in this paper the results of my study with applications to arithmetic.[32]

He then devotes a rather substantial preface to the logical and arithmetical notation he has developed – a notation which approximates the symbolic languages of today in precision. For our purposes, it is perhaps most relevant to notice that Peano distinguishes, for the first time, the membership relation, $\in$, from the subset relation. $\subseteq$.[33]

The main text of Peano's work is entirely symbolic and shows how his new notation can be applied to the derivation of arithmetical theorems. The structure of this derivation is spelled out in Peano's preface:

> Propositions which are deduced from others by the operations of logic are *theorems;* those for which this is not true I have called *axioms.* There are nine axioms here ($\P$1), and they express fundamental properties of the undefined signs.
>
> In $\P$1–$\P$6 I have proved the ordinary properties of numbers.
> . . .
>
> In the following sections I have treated various things so that the power of the method is better seen. In $\P$7 are several

---

[31]See [**75**, pp. 101–134]. Peano's booklet cites the two articles by Peirce, [**78**] and [**80**], but does not cite Peirce's 1881 paper. See [**53**, p. 102n]. And Peano cites Dedekind in the text. "Also quite useful to me was the recent work by R. Dedekind, *Was sind und was sollen die Zahlen?*, in which questions pertaining to the foundations of numbers are acutely examined." *Ibid.*, p. 103.

[32]*Ibid.*, p. 101

[33]*Ibid.*, pp. 102, 107-108. Peano uses the same symbol for set inclusion as for logical implication, but these always distinguished by context.

theorems pertaining to the theory of numbers, In ¶8 and ¶9 are found the definitions of rationals and irrationals. Finally, in ¶10 I have given several theorems, which I believe to be new, pertaining to the theory of those entities which Professor Cantor has called *Punktmenge*.[34]

It is evident that Peano's theorems extend further into the body of arithmetic than those of Dedekind and Peirce. He not only treats the integers, rationals, and reals, but also derives a portion of traditional number theory and proves several theorems concerning point sets.

But the achievement for which Peano is best known is the compact axiom system from which all of these theorems are deduced. Because of their economy and power, the Peano axioms – also referred to as the "Peano postulates" – have become a staple of modern foundations.[35]

In Peano's original formulation there were nine axioms with four undefined terms. With two minor changes in notation, adopting the separate symbol, $\Rightarrow$, for logical implication *and* replacing Peano's system of dots with the modern use of parentheses over the hierarchy $\Rightarrow, \&, =, \in, +$, these are as given in the first section of *Arithmetices principia*:

Explanations

The sign N Means *number* (*positive integer*); 1 means *unity*; $a + 1$ Means the *successor of* a, or a *plus* 1; and $=$ means *is equal to* (this must be considered as a new sign, although it has the appearance of a sign of logic).

Axioms

1. $1 \in N$
2. $a \in N \Rightarrow a = a$
3. $a, b \in N \Rightarrow (a = b) = (b = a)$
4. $a, b, c \in N \Rightarrow (a = b \,\&\, b = c \Rightarrow a = c)$
5. $a = b \,\&\, b \in N \Rightarrow a \in N$
6. $a \in N \Rightarrow a + 1 \in N$
7. $a, b \in N \Rightarrow (a = b) = (a + 1 = b + 1)$
8. $a \in N \Rightarrow a + 1 \neq 1$
9. $k \in K \,\&\, 1 \in k \,\&\, (x \in N \,\&\, x \in k \xrightarrow{x} x + 1 \in k) \Rightarrow$
$$N \subseteq k.[36]$$

---

[34]*Ibid.,*, p. 102.

[35]For an excellent informal explanation of the power of the Peano axioms, [**46**, pp. 372ff.].

[36]See [**53**, p. 113].

Notice that in Peano's ninth (mathematical induction axiom, "$k \in K$" should be read as "k is a class", and universal quantification is indicated by placing the variable, x, under the implication sign. Despite Peano's caution not to confuse it with a logical sign, the properties of equality (axioms 2-.5) are generally given today as part of the underlying logic.[37]

This allows a quite simple statement of Peano's system as five axioms containing three undefined terms.[38] In a common later formulation, using the primitives "1," "number," and "successor," the Peano axioms are:

1. 1 is a number
2. The successor of any number is a number
3. No two numbers have the same successor
4. 1 is not the successor of any number
5. Any property which belongs to 1, and also to the successor of every number which has the property, belongs to all numbers.[39]

From an historical point of view, it is unclear to what extent Peano's axioms were actually derived from the simply infinite system of Dedekind.[40] But the resemblance between the two systems is apparent. Peano's second axiom is equivalent to asserting that numbers form a chain with respect to the successor function; his third axiom says that such a function must be one-to-one; and his fourth axiom prohibits 1 from being an element of the image of the set of numbers. And although they are not precisely equivalent, Peano's mathematical induction axiom corresponds closely to Dedekind's equation of the set of numbers with $\{1\}_\circ$, the intersection of all those chains which contain

---

[37]The reason for Peano's caution. as can be seen from axiom seven, is that he uses the same symbol for equality as for the logical biconditional. On methods of defining equality, see [**97**, 12–44].

[38]Hao Wang points out that "it is possible to absorb the concept "number" and the first two axioms into an explicit specification of notation," leaving just three axioms and two undefined terms[**123**, p. 149]. And Bertrand Russell suggests defining "number" after the manner of Dedekind, as "the posterity of 1 with respect to the relation 'immediate predecessor'," thus eliminating axioms one and five. [**112**, pp. 22–23].

[39]Adapted from [**112**, pp. 5–6].

[40]A direct historical dependence is claimed by Wang, [**123**, p. 149], and is implied by [**10**, p. 126] But such a dependence is denied by Kennedy [**53**, p. 101] as well as by, e.g., Morris Kline, [**55**, p. 988].

1 as an element.[41] Finally, the first Peano axiom is needed only to guarantee that the set of numbers will not be empty, an eventuality which Dedekind had rejected on principle.[42] The overall equivalence of the Peano axioms and the Dedekind axioms has long been general knowledge.[43] For this reason, and because Dedekind's essay contains other interesting points of comparison with Peirce, the rest of the present chapter will deal primarily with the approaches of Peirce and Dedekind.

1.4. Transitive Relations and Functions

Peirce's axiom system for the natural numbers provides a basic contrast to the systems of Dedekind and Peano: it is based upon a transitive relation instead of a successor function. In this section, we shall try to gain some understanding of what this difference means.

What is a relation? Although Peirce was largely responsible for initially developing the logic of relations, and defined transitive relations as early as 1870, he used the notion of a relation more as what we would today call a "predicate."[44] Similarly. Dedekind originally defined a function as "a law according to which to each definite element s of S a definite thing belongs" (§21). In both cases one is left with the typical dilemma of intensions, viz., how to determine when two predicates or laws are identical. For mathematical purposes what is required is a completely extensional definition of these notions, so that two relations or functions are considered identical if and only if they relate exactly the same elements. Such a definition was not given until 1921 by Casimir Kuratowski.[45] Kuratowski defined a (binary) *relation* as a set of ordered pairs, which were in turn defined as particular sets of sets.[46] Since x

---

[41]They are not precisely equivalent because, e.g., the Dedekind axiom $N = \{1\}_o$ implies that N is a chain while the Peano mathematical induction axiom does not have the corresponding implication that the successor of any number is a number. (Peano's axioms, as stated. are mutually independent but Dedekind's first axiom can be derived from his second). On the similarity between the two axioms, see further below, pp. 30, 31.

[42]Given only that the set of numbers is not empty, Peano's first axiom can be derived from his fifth. It is clear that Dedekind intended 1 to be an element of N, since he spoke of it as "the base element of N," and his third axiom stated that "the element 1 is not contained in N'," which, as the sequel make evident, suggests that 1 is contained in N. But Dedekind conceived $1 \in N$ to be derivable from $N = [1]_o$ via theorem §45, which is only the case if N is non-empty. The formalization in Chapter 2 will explicitly conjoin $1 \in N$ to Dedekind's second axiom, see below, pp. 75, 76.

[43]See the comparison in [**10**, pp. 113–131].

[44]3.136. Also see 3.63. 3.218. 3.456–3.487.

[45]See [**58**].

[46]E.g., $(x, y) = \{\{x\}, \{x, y\}\}$. Specific definitions of these terms are given in Chapter 2.

is related to y just in case the ordered pair (x, y) is an element of the set of similarly related pairs, this definition allowed the extensionality of relations to be derived directly from the extensionality of sets. And because functions can be defined as a kind of relation, it made possible the reduction of the entire mathematical theory of relations and functions to the theory of sets.

Until the twentieth century, the mathematical conception of a function was somewhat amorphous[47] But even without the set-theoretical reduction of relations, the notion of a function can be clarified by interpretation as a kind of relation instead of some vaguely conceived "law."[48] Thus a function, f, can be described as a relation in which each element on the left (the domain) is related to but a single element on the right (the range). We will speak of a *function on a set N* when N is identical to the domain of the function. An obvious example is the relation "is immediately succeeded by" which is a function on the set of natural numbers.[49]

Like functions, transitive relations are just another species of relation. We will call a relation *transitive in a set N* when for any elements, x, y, and z, in N, if x is related to y and y is related to z, then x is related to z. (A relation is simply *transitive* when it is transitive in its field, i.e., throughout its range and domain taken together.) The obvious examples are the relations "greater than" and "greater than or equal to" which are transitive in the natural numbers.

It is easy to see that normally a relation will not be both transitive and functional. In particular, it is clear that Dedekind's successor function is not transitive, and that Peirce's transitive relation is not functional. This raises two important questions, 1) How are the axiom systems of Peirce and Dedekind to be compared if they are based upon such unlike primitives? and 2) Are there any reasons to prefer the use of either primitive over the other? We will examine these in order.

In regards to the first question it should be noted that the primitives of Peirce and Dedekind can be cross-defined. Peirce, for instance, defined the relation "next greater than," the inverse of which is Dedekind's successor function, entirely in terms of his transitive primitive q. And Dedekind defined

---

[47]See [**63**, pp. 69–76].

[48]Neither Dedekind nor Peano views a function as a kind of relation. But Peirce later attributed the first step in this direction to Dedekind's 1879 edition of Direchlet's *Vorlesuyyen über Zahlentheorie* which generalizes the notion to discrete sets, and credited Ernst Schröder with explicitly defining functions as relations in 1895. See 3.610.

[49]Although we will speak of this as the "successor" function, we will generally describe functions with the domain on the left. There is a tradition, defended in Quine [**97**, pp. 24–36], of doing the opposite.

the transitive relation "greater than or equal to" entirely in terms of his own successor function (§93). Peirce's definition is rather straightforward and has already been given above. Dedekind's definition says that if x and y are natural numbers, x is greater than or equal to y just in case x belongs to every chain to which y belongs. Dedekind's definition was of some historical importance, having been first formulated by Frege in 1879 to characterize what he called the "ancestral" of a function.[50]

If the primitives of Peirce and Dedekind can be cross-defined, the first question is almost answered. But it is important not to suppose that the possibility of a definition can substitute for the definition itself. In order to demonstrate the equivalence of the axiom systems of Peirce and Dedekind, it will be necessary to explicitly conjoin these definitions to their axiom systems. This will extend the semantics of both systems – instead of the triplets $< N, R, 1 >$ and $< N, f, 1 >$, models for both will consist of a quadruplet $< N, R, f, 1 > -$ without, however, affecting the syntactical strength of either.

In response to the second question, it will be helpful to look briefly at the notion of partial order. We say that a relation is *antisymmetric in a set N* when for all elements, x and y, in N, if x is related to y and y is related to x, then x is equal to y. And a relation is called *reflexive in a set N* when every element in N is related to itself. The relation "greater than or equal to" is both antisymmetric and reflexive in the set of natural numbers.

Corresponding to the weak relation "greater than or equal to" there is a strict relation "greater than" which, while transitive, is neither antisymmetric nor reflexive. We will call a relation *asymmetric in a set N* when for all elements, x and y, in N, if x is related to y then y is not related to x. And we will say that a relation is *irreflexive in a set N* when no element in N is related to itself. In the presence of transitivity these latter two properties are equivalent, hence the strict relation "greater than" can be described indifferently as being transitive asymmetric or transitive irreflexive in the natural numbers. The notion of partial order can be introduced via either a weak or a strict relation. Choosing the former, for example, one would say that a set N is (weakly) *partially ordered by a relation R* when R is transitive, antisymmetric. and reflexive in N. It is clear that Peirce's system of quantity is partially

---

[50]See [**32**, pp. 55ff.]. In an 1890 letter to H. Keferstein, Dedekind remarked that; "For a brief period last summer Frege's *Begriffsschrift* . . . came into my possession for the first time. I noted with pleasure that his method of defining a relation between an element and another which it follows in a sequence, not necessarily immediately, agrees in essence with my concept of chains." In explaining the importance of Frege's definition Bertrand Russell says, "until Frege developed his generalized theory of induction, no one could have defined 'ancestor' precisely in terms of 'parent' ." See [**112**, pp. 25–27, 35, 43–46].

ordered. since his fundamental relation of quantity, q, is defined as having precisely the three requisite properties. The most common example of such a system is the partial ordering produced by the subset relation in reference to any collection of sets.[51]

The philosophical importance of partially ordered sets is brought out in Bertrand Russell's *Principles of Mathematics*. Russell suggests that such sets contain the minimum order which allows "betweenness" to be defined. "Any order at all," Russell says, "implies the presence of a transitive asymmetric relation."[52] Thus when Russell compares Dedekind's method of generating numbers through a successor function with Peirce's approach using a transitive relation, he concludes that the latter is to be preferred because it expresses the more fundamental order properties of such series.[53]

On this point Russell would seem to be correct. The natural ordering of many sets, e.g., the set of fractions, the real numbers, the moments in time, etc., is incompatible with a successor function. Yet all such sets have in common the fact that they are partially ordered by a transitive relation. In this respect, then, Peirce's transitive primitive has the advantage of being conceptually more general than Dedekind' s successor function. Peirce himself seems to have realized this since he later described transitive relations as "the basis of all quantitative thought."[54]

In answer to our second question, then, the functional approach, insofar as it yields the Peano axioms, is the more facile technically and has an obvious edge in simplicity. But Peirce's transitive approach has the philosophical advantage of allowing him to describe the natural numbers *genus et differentia*, as one particular species of order.

---

[51]See, e.g., Dedekind's theorems, §4, §5, and §7.

[52]See [**109**, pp. 199–217, especially p. 217] and [**112**, pp. 29–41, especially pp. 29–30].

[53]See [**109**, p. 217 and pp. 243–244].

[54]4.94. There is an actual historical link between Peirce and Russell on this topic. Russell initially emphasized partial order in his paper "On the Notion of Order" [**108**] claiming that "such series have great philosophical importance," and that their theory "is one of the most essential parts of logic." But at the beginning of his paper Russell pointed out that "the following account of the genesis of order is virtually identical with that of Mr. B. I. Gilman." The work to which Russell referred was an 1892 paper by Benjamin Ives Gilman, "On the Properties of a One-Dimensional Manifold," [**35**] in which partial ordering was held to exhibit "the ultimate constituents of the notion of one-dimensionality." Gilman, though, was a student of Peirce's at Johns Hopkins at the time "On the Logic of Number" was written. Gilman cited Peirce, and there can be little question but that his work was strongly influenced by Peirce. It is interesting, in this connection, to notice that Russell always cites Peirce's discovery of irreflexive (aliorelative) relations. E.g., see [**109**, p. 203] and [**112**, p. 32]. Peirce has not generally been recognized for the important role he played in founding the theory of partial order – an important topic in modern set theory.

Depending upon whether one begins with a transitive relation or a function, one's description of the natural numbers will follow different courses. But regardless of the starting point and subsequent path adopted, all such descriptions have a common destination. There are certain general tasks which must be performed before we would call any axiomatization successful. The next two sections will introduce some of these tasks, exhibiting some common themes in the otherwise different axiom systems of Peirce and Dedekind. Chapter 2 will complete this comparison by showing that the two axiom systems are, in fact, equivalent.

1.5. Simple Order and Chains

A set N is said to be *connected by a relation R* when for all elements, x and y, in N, either x is related to y or y is related to x. One would clearly want the natural numbers, for example, to be connected by the relation "greater than or equal to."

A set N is said to be *simply ordered by a relation R* when N is both partially ordered and connected by R. Thus Peirce's "system of simple quantity" merely defines a simply ordered set, since it adds to the description of quantity (partial order) the further requirement of connection.[55] Although modern set theory has spawned a superfluity of names for such sets, e.g., "ordered sets," "linearly ordered sets," "totally ordered sets ," for our purposes it is most relevant to notice that simply ordered sets are occasionally referred to as "chains."[56] More specifically, the terms "chain" and "monotone set" are often used to designate a set which is simply ordered by the subset relation.[57] Now this is *not* how Dedekind uses the term. Nonetheless, there is a natural relation between Dedekind's conception of a "chain" and the modern notion of a "monotone set."

Let P be a chain in the Dedekind sense, so that in respect to some function f, $P' \subseteq P$. Now consider any set composed entirely of images of P gotten by iteration of f, i.e., any subset of the sequence $\{P, P', P'', \ldots\}$. Like any family of sets, such a set will be partially ordered by the subset relation. In addition, since P is a chain, it is intuitively evident that every element of such a set will

---

[55]Although it is a striking coincidence, the term "simple" seems to have entered modern usage through Cantor, *einfach geordnete Menge* [**18**, p. 110], rather than through Peirce. See also [**52**, p. 10]. It is clear that Peirce initiated the study of simple order independently of Cantor.

[56]See [**10**, p. 54].

[57]See [**117**, p. 244] and [**31**, p. 128]. A Dedekindian use of the term can still be found in [**61**, p. 126].

either include or be included in every other element of the set.[58] It follows that the set will be connected, and thus simply ordered, by the subset relation. Hence, it will be a monotone set or "chain" in the modern sense.

Aside from illustrating how the term "chain" has changed from Dedekind's original understanding of it, the notion of a monotone set is important because it characterizes Dedekind's own method of ordering the natural numbers. Given merely the set of numbers $N = \{1, 2, 3, \ldots\}$ and a successor function according to which these numbers constitute a chain, one can produce a simple ordering by constructing the series of images obtainable from this set, i. e. , the family of sets:

$$\{1, 2, 3, 4, 5, \ldots\}$$
$$\{2, 3, 4, 5, \ldots\}$$
$$\{3, 4, 5, \ldots\}$$
etc.

This family of sets does appear to comprise a monotone set, simply ordered by inclusion. Dedekind's procedure, then, is to identify each natural number with the appropriate image, thus ordering the set of numbers by reference to the order already present in the set of images.[59] The problem, of course, is to guarantee first of all that the set of images does look like the series of remainders given above, and secondly that each image can in fact be uniquely corresponded with a natural number.

Although Dedekind's treatment of ancestrals, i.e., of *the* chain of a given set, has the broader function of establishing the inductive character of the natural numbers, it is worth noticing how it serves to correlate each number with its appropriate remainder set. We have already mentioned that the chain of a set of natural numbers with respect to the successor function will consist of all natural numbers greater than or equal to the minimum number in that set. Thus $\{n\}_\circ$, the chain of the unit set $\{n\}$, will include all natural numbers greater than or equal to n. So *the* chain of each natural number (taken as a unit set) will provide the remainder with which that number is associated. It is not difficult, in fact, to prove that the collection of chains thus obtained is connected, i.e., that for all numbers, x and y, either $\{x\}_\circ \subseteq \{y\}_\circ$ or $\{y\}_\circ \subseteq \{x\}_\circ$.[60] It

---

[58]The proof of this, as well as the definition of the set itself, requires mathematical induction.

[59]This technique has the economy of ordering the natural numbers by reference to the primitive relation of the underlying set theory. Notice that Dedekind's approach requires infinite sets. These could be avoided by identifying numbers with initials rather than remainders. See Chapter 3.

[60]See Chapter 2, proof P2.

follows that such a collection of chains will be simply ordered by the subset relation. And the ordering of the natural numbers themselves by reference to this order is straightforward: for any natural numbers, x and y, $x \geq y$ just in case $\{x\}_\circ \subseteq \{y\}_\circ$ or, what is the same, $\{x\} \subseteq \{y\}_\circ$. It is important, however, not to conclude that the natural numbers are simply ordered by the relation "$\geq$" until one has shown that the correspondence between numbers and remainders is one-to-one. In particular, one must show that for all numbers, x and y, $\{x\}_\circ = \{y\}_\circ \Rightarrow x = y$ before one can assume that antisymmetry will transfer from the set of remainders to the set of natural numbers.[61]

Having defined a set which is simply ordered, one has not yet described the natural numbers. The set of positive fractions, for instance, is simply ordered by the relation "greater than or equal to" but is not similar to the natural numbers. So, among other things, it is necessary to further confine simple order by ruling out systems which are *dense*, i.e., infinitely divisible. Dedekind, by choosing a successor function as primitive and then stipulating the closure of succession in his first axiom, does bypass those systems which, like the positive fractions, are everywhere dense. For Peirce, though, such systems are not ruled out until his third axiom provides closure with respect to immediate predecessors. And even so, both Peirce and Dedekind need to further insure that their numbers are not dense in specially selected regions. Another way of putting this is to note that simply ordered sets may be "too large" for the natural numbers, and need to have their ordinal type restricted so as to guarantee that no initial segment of natural numbers is infinite. This "minimalization" is the topic of our next section, on mathematical induction.

Finally, it should. be mentioned that simply ordered sets may also encompass the opposite extreme of being "too small" for the natural numbers, i.e., such sets may be finite. Dedekind's first, third, and fourth axioms together guarantee that his numbers will in fact be infinite, while Peirce's fourth axiom (given simple ordering) provides a similar guarantee for his numbers.

1.6. Mathematical Induction

The principle of mathematical induction lies at the heart of the natural number system, and it is virtually impossible to prove anything in elementary arithmetic without this principle. Although it can be formulated in many different ways, it is most often stated after the fashion of our fifth Peano axiom:

---

[61]See Chapter 2, lemma L3.

> Any property which belongs to 1, and also to the successor of
> every number which has the property, belongs to all numbers.

It is important not to confuse this principle, sometimes referred to as *weak
mathematical induction,* with a different principle known as *transfinite induction*:

> If any number has a property whenever all numbers less than
> that number have the property, then the property belongs to
> all numbers.

The latter principle supplies a method of inference applicable to sets of a higher
ordinal type than the natural numbers (in fact, to any well ordered set). Weak
mathematical induction, on the other hand, is exclusively characteristic of systems similar to the natural numbers or finite. We are primarily interested
in this weaker principle, and so will henceforth understand it by the phrase
"mathematical induction."[62]

In his later writings, Peirce generally referred to mathematical induction
by the name "Fermatian inference." Aside from holding that the principle
derived from the method of "infinite descent" in Fermat, Peirce was concerned
that it not be confused with ordinary (scientific) induction which it resembles
only superficially.[63] Peirce would probably have agreed with Bertrand Russell's
response to the question "Why does mathematical induction work?" Russell

---

[62]There is a third formulation of mathematical induction which is sometimes called
course of values induction:

> Any property which belongs to 1, and belongs to the successor of a number
> whenever it belongs to every number less than or equal to that number,
> belongs to all numbers.

This form of induction is equivalent to weak mathematical induction and can be interchanged
with it in any context. The only difference is that the inductive step proceeds from the set
of numbers between 1 and n to $n+1$, rather than from n to $n+1$ as in weak mathematical
induction. This difference is merely a matter of bookkeeping.

Transfinite induction differs from both weak induction and course of values induction in
that its inductive step does not require that there be a relation of succession between n and
any of the set of numbers less than n. In fact, there is no mention of succession at all in the
formulation of transfinite induction. This means that transfinite induction can be applied in
situations in which certain elements lack predecessors.

Unfortunately, because of their superficial similarities, elementary texts sometimes refer
to both course of values and transfinite induction as *strong* mathematical induction. This
appellation would be more suitably reserved for the latter. Given the natural numbers all
three versions work. But only weak mathematical induction and course of values induction
can serve the purpose of axiomatization.

[63]See [28].

pointed out that mathematical induction is simply *defined* to work for the natural numbers:

> If 'quadrupeds' are defined as animals having four legs, it will follow that animals that have four legs are quadrupeds; and the case of numbers that obey mathematical induction is exactly similar.[64]

It is true, of course, that having described the natural numbers one expects this description to conform roughly to the intuitions of mathematicians. And such an observation is likely behind Poincaré's contention that mathematical induction is a synthetic principle incapable of justification except by intuition. But Poincaré was really concerned with the broader question of Russell's logicism, rather than with the principle of mathematical induction itself. His arguments are directed against any axiomatization in the service of logical reduction, alighting on mathematical induction as a sort of test case.[65] Peirce, in fact, would be sympathetic to Poincaré's general position, and would probably accept his argument to the extent of granting that intuition is operative in one's initial formulation of mathematical induction as a hypothesis. Given his system of "ordinary number," however, Peirce would view mathematical induction as a necessary consequence of the fifth axiom, and would agree with Russell in claiming that the conclusions obtained by the application of induction to such a system will have the same force as any deductive inference.[66]

Peirce's definition of an "infinite class," cited at the beginning of this chapter, is one of the four different formulations of mathematical induction given in his paper (3.258). We refer to it as "mathematical induction starting with k," since it differs somewhat from the usual Peano formulation:

> Any property which belongs to a number k, and also to the successor of every number which has the property, belongs to all numbers greater than k.

In the presence of the other (Peirce or Peano) axioms, mathematical induction starting with k is equivalent to the Peano mathematical induction axiom. Letting $k = 1$, it obviously implies the Peano version, while by substituting properties of the form "less than k or having the property P" it can be seen to be implied by the Peano axiom. It should be noted, though, that taken in isolation the two formulations are not equivalent. Induction starting with k,

---

[64]See [**112**, p. 27].

[65]See [**88**, ch. 4–5] and [**10**, pp. 69, 70, 113–123].

[66]Peirce's position will be more fully developed in Chapter 3.

for instance, is characteristic of the system of integers, while the more common formulation is not.[67]

We have been stating the principle of mathematical induction in terms of "properties" but we could also state it extensionally in terms of the "classes" or "sets" of elements having such properties. Since the full mathematical induction principle refers to "any" or "all" such properties or classes, it is generally taken to require a second-order predicate logic – a logic capable of quantifying over predicates or classes. We will return to this issue in the next section, but for now it is important to note that the axiom systems of Peirce, Dedekind, and Peano, all have full mathematical induction axioms.

Peano and Peirce introduce the principle of mathematical induction as an axiom. Dedekind, on the other hand, derives this principle as a theorem. That this difference is logically inconsequential was pointed out by Bertrand Russell:

> As regards [Dedekind's] proof of mathematical induction, it is to be observed that it makes the practically equivalent assumption that numbers form the chain of one of them. Either can be deduced from the other, and the choice as to which is to be an axiom, which a theorem, is mainly a matter of taste.[68]

Dedekind's second axiom, $N = \{1\}_\circ$, is not strictly equivalent to the Peirce/Peano mathematical induction axioms, but one can say with Bertrand Russell that they are "practically equivalent", that they make the same substantive contributions to their respective axiom systems.[69] Consider, for instance, the Peano induction axiom. The inductive hypothesis supposes that whenever a property belongs to a number, it also belongs to the successor of that number. But this is the same as saying that in respect to the successor function, such properties – construed extensionally as sets – constitute *chains*, i.e., if P is such a property, we must have $P' \subseteq P$. Taken in this sense, the entire principle would then assert that if i) 1 belongs to such a set P, and ii) P is a chain, then all numbers belong to such a set P. Since this is stated to be the case for any P, it follows that numbers belong to every chain containing the number 1, and thus that numbers belong to the *intersection* of all those chains containing the number 1. Now if one makes the additional assumption that numbers themselves constitute a chain with respect to the successor function,

---

[67]This is probably why Peirce chose this formulation – in order to simplify the transition to an "unlimited" system.

[68]See [**109**, p. 248].

[69]They are not strictly equivalent since, taken in isolation, the Dedekind axiom implies that N is a chain. See above, p. 19.

which is Dedekind's first axiom and Peano's second axiom, it would have to be the case that numbers are exactly the intersection of all those chains containing the number 1. And this is simply Dedekind's second axiom, that $N = \{1\}_\circ$.

The derivation in the other direction is to be found entirely within Dedekind's essay, and differs little in conception from the correlation outlined above. Dedekind's approach, though, is noteworthy because of its technical elegance. He first establishes the general "theorem of complete induction": given the function f and two sets, A and P, if i) $A \subseteq P$, and ii) $(A_\circ \cap P)' \subseteq P$, then $A_\circ \subseteq P$. This theorem, which remarkably holds for any sets and regardless of whether f is one-to-one, is a consequence of the definition of $A_\circ$. This can be seen by considering the theorem (§47), $A \subseteq K$ & $K' \subseteq K \Rightarrow A_\circ \subseteq K$, and substituting the set $A_\circ \cap P$ for the set K.[70] Now if we further substitute the unit set $\{1\}$ for A and take as given the axiom $N = \{1\}_\circ$, the theorem of complete induction implies that if i) $\{1\} \subseteq P$, and ii) $(N \cap P)' \subseteq P$, then $N \subseteq P$. But if f is a successor function, this will clearly be translated into property-language by the statement: If 1 has a certain property, and whenever a number has that property its successor also has that property, then all numbers must have the given property – which is the principle of mathematical induction.[71]

This kinship to Dedekind's second axiom helps explain why the principle of mathematical induction is considered essential in axiomatizing the natural numbers. It is generally agreed, to begin with, that the natural numbers must form a chain with respect to the successor function. (This is also expressed by saying that they are "closed" or that the property of being a natural number is "hereditary" with respect to succession.) Now if it is stipulated that such a chain be infinite, it follows that it must be at least large enough for the natural numbers. The problem, though, is that such a chain may include elements beyond those desired, e.g., like the series of fractions,

$$-1, -\frac{1}{2}, -\frac{1}{4}, \ldots ; 1, \frac{1}{2}, \frac{1}{4}, \ldots$$

which is a chain with respect to the function of halving. This chain clearly includes more, structurally, than the plain "one damn thing after another" that we want for the natural numbers. The last part of the series cannot even be reached from its beginning in a finite sequence of steps (in fact, the stipulation that all elements of a chain should be reachable in a finite number of steps is equivalent to mathematical induction).

---

[70]A proof of the theorem of complete induction is given in Chapter 2, theorem T11.

[71]Dedekind interprets the theorem of complete induction (§59) in terms of properties at (§60), and derives the Peirce and Peano axioms at (§80).

In order to exclude such unwanted "extra" elements, Peano and Dedekind consider all possible chains, or hereditary properties, with respect to the successor function and, in effect, describe the natural numbers as the "minimal" such chain containing the initial element of the series. This is what it means to assert that numbers belong to the intersection of all chains containing the number 1. Mathematical induction, then, serves to prohibit succession from capturing unwanted elements external to the system of natural numbers. Dedekind, in fact, emphasizes this same point in an 1890 letter to H. Keferstein:

> But I have shown in my reply that these [other axioms] are still far from being adequate for completely characterizing the nature of the number sequence N. All these [other axioms] would hold also for every system S that, besides the number sequence N, contained a system T, of arbitrary additional elements t, to which the mapping ϕ could always be extended while remaining similar [one-to-one] and satisfying $\phi(T) = T$. But such a system S is obviously something quite different from our number sequence N, and I could so choose it that scarcely a single theorem of arithmetic would be preserved in it. What, then, must we add to the facts above in order to cleanse our system S again of such alien intruders t as disturb every vestige of order and to restrict it to N? This was one of the most difficult points of my analysis and its mastery required lengthy reflection. If one presupposes knowledge of the sequence N of natural numbers and, accordingly, allows himself the use of the language of arithmetic, then, of course, he has an easy time of it. He need only say: an element n belongs to the sequence N if and only if, starting with the element 1 and counting on and on steadfastly ... I actually reach the element n at same time; by this procedure, however, I shall never reach an element t outside of the sequence N. But this way of characterizing the distinction between those elements t that are to be ejected from S and those elements n that alone are to remain is surely quite useless for our purpose; it would, after all, contain the most pernicious and obvious kind of vicious circle ... Thus, how can I, without presupposing any arithmetic knowledge, give an unambiguous conceptual foundation to the distinction between the elements n and the elements t? Merely through consideration of the chains (§37 and §44 of my essay), and yet, by means of these, completely: If I wanted to avoid my technical expression "chain" I would say: an element n of S belongs to the sequence N if and only if n is an element of *every* subset K of

S that possesses the following two properties. (i) the element 1 belongs to K and (ii) the image $\phi(K)$ is a subset of K. In my technical language. N is the intersection $\{1\}_\circ$ of all those chains K (in S) to which the element 1 belongs. Only now is the sequence N characterized completely.[72]

Notice that Dedekind considers the possibility of a set S which, in addition to the set N of natural numbers, contains an arbitrary set T of unwanted elements. Provided only that the function f is extended in such a way as to remain one-to-one while giving $T' = T$, all of Dedekind's other axioms will be satisfied by the set S, yet the set S will not be similar to the natural numbers since certain properties provable only by using mathematical induction will not hold for it. For example, Dedekind allows that T might be *finite* and in the limit case, where T consists of but a single element, such an element would have to be its own successor. So the law, provable for N, that no element can be its own successor, has a clear counterexample in the case of such sets S.

With Peirce's approach, on the other hand, an element which is its own successor is ruled out by definition. The immediate successor of an element is simply defined as being greater than but not equal to that element. In fact, it would be possible to show that no *finite* set T of unwanted elements could be added so as to remain consistent with the rest of Peirce's axioms. Intuitively, the reason for this is that such a set T would itself have to be simply ordered and, since it is finite, have a relative maximum element t_m. But this element t_m, by the requirement of connection, would either be greater than every natural number, or less than or equal to some particular natural number n. In the former case t_m would violate Peirce's fourth axiom since it would be a maximum for the entire set S; while in the latter case every element of T would be less than or equal to the number n and it would be possible to show that the addition of T has in no way changed the properties belonging to N, that S is similar to N, and thus that T is not really unwanted.[73] Hence, unlike the systems of Dedekind and Peano, Peirce's other axioms do not necessarily permit the addition of an *arbitrary* set T of unwanted elements. This does not mean, however, that it is not possible to choose a set T of unwanted elements which could be added to the natural numbers in such a way as to still satisfy Peirce's other axioms. It only means that such a set T cannot be finite.[74]

---

[72]Dedekind's letter of February 27, 1890, is translated by Stefan Bauer-Mengelberg after Hao Wang, in [**43**, 98–103].

[73]Our conception of what constitutes an "unwanted" element is, admittedly, intuitive. Any attempt to provide a formal criterion would be circular, in that, if successful, it would ultimately reduce to the principle of mathematical induction. Such is the case, for example, with defining an element as "unwanted" when it corresponds to an infinite initial segment.

[74]We are using the terms, "finite" and "infinite," in their ordinary non-Dedekind sense.

All three approaches, in fact, depend upon the principle of mathematical induction to exclude from the natural numbers unwanted sets T which are infinite. To help see this, we will introduce the notion of well ordering. A set N is said to be *well ordered by a relation* R when N is partially ordered by R and every non-empty subset of N contains a least element. (It follows that every such set will also be simply ordered, since of any two elements one must be "least," thus insuring connection.) The series of fractions given above would be just one example of a set which is well ordered.[75] In general, such sets can have the form,

$$S = \{a_1, a_2, a_3, \ldots ; b_1, b_2, b_3, \ldots ; c_1, c_2, c_3, \ldots ; \text{etc.}\}$$

containing, in principle, an infinite sequence of unwanted infinite series. Any sequence S of such infinite series, i.e., any well ordered set without a maximum, will satisfy the other (non-inductive) axioms of Dedekind and Peano. It is clear, for instance, that a successor function defined in the expected way will be one-to-one, that the least element, "a_1" will belong to S but not be a successor, and that S will be a chain with respect to succession (if x is an element of S, then its successor $f(x)$ is also an element of S). And yet most sequences S, such as our series of fractions, will not be similar to the natural number system since they include one or more unwanted infinite sets T. The mathematical induction axiom, by describing the natural numbers as the *minimal* set which satisfies the other (non-inductive) conditions, in effect restricts any such sequence S to its initial infinite segment $\{a_1, a_2, a_3, \ldots\}$. In this respect, mathematical induction functions as a *minimalization principle* for well ordered sets.

With Peirce, the possibilities compatible with his initial axioms are clearly different. As we have shown, Peirce's initial axioms rule out the possibility of an arbitrary finite set T of unwanted elements. And in case T is infinite, sets which are well ordered, such as our series of fractions, are prohibited by Peirce's third axiom requiring closure with respect to immediate predecessors. In the well ordered set S, for example, the elements b_1 and c_1 do not have immediate predecessors in S. But there are other possibilities, containing unwanted infinite sets, which are compatible with Peirce's initial axioms. Such sets might have the form:

$$S_1 = \{a_1, a_2, a_3, \ldots ; , \ldots, b_3, b_2, b_1; c_1, c_2, c_3, \ldots\}$$

and, in general, could contain one or more infinitely descending series

$$\{\ldots, b_3, b_2, b_1\}.$$

---

[75]Given the usual ordering of the rationals, it is easy to define an ordering relation for this set. E.g x is related to y just in case y is negative and less than x or y is positive and greater than x (for strict ordering). In general, when we exhibit sets informally. as ordered, we assume that the visual ordering from left to right can be given precise definition.

For Peirce, the mathematical induction axiom is needed in order to restrict sets such as S_1 to systems similar to the natural numbers. It speaks for Peirce's thorough understanding of mathematical induction that he saw that the principle would serve to simultaneously minimalize and well order such sets. The proof that Peirce's system of "ordinary numbers" is actually well ordered constitutes perhaps the least obvious result of the equivalence proof in Chapter 2.[76] But despite this difference, it is evident that Peirce, like Dedekind, conceived the axiomatic role of mathematical induction to be primarily that of minimalization. This is why he termed classes obeying mathematical induction "infinite" – in order to distinguish them from their "superinfinite" alternatives. In general, it is not difficult to understand the axiomatic role of minimalization by reference to our intuitive expectations of what the natural numbers should look like. The significance of a minimalization principle can be expressed in several ways: it insures, for instance, that the natural numbers will be comprised of elements which can be reached by counting "on and on steadfastly," i.e., that every initial segment of natural numbers will be finite; and it restricts the "ordinal type" of the natural numbers by excluding those sets of transfinite numbers which would otherwise satisfy the axioms. Mathematical induction does have a further role, perhaps best termed "meta-axiomatic," in establishing the categoricity of axiom systems for the natural numbers – a topic we will examine in more detail below.

Peano and Dedekind are more similar in regard to the axiomatic function of mathematical induction in their systems. But Peirce and Peano are more similar in explicitly stating the inductive method of inference as an axiom rather than a theorem. Concerning the latter point, Bertrand Russell prefers the procedure of Peano and Peirce. He argues that:

> On the whole, though the consideration of chains is most ingenious, it is somewhat difficult, and has the disadvantage that theorems concerning the finite class of numbers not greater than n as a rule have to be deduced from corresponding theorems concerning the infinite class of numbers greater than n. For these reasons, and not because of any logical superiority, it seems simpler to begin with mathematical induction. And it should be observed that, in Peano's method, it is only when theorems are to be proved concerning any number that mathematical induction is required ...In Dedekind's method, on the other hand, propositions concerning particular numbers, like general propositions, demand the consideration of chains.

---

[76]The most important part of this proof, that Peirce's numbers form a chain with respect to the successor function, is the second half of the proof of D0 on pp. 90–92.

Thus there is, in Peano's method, a distinct advantage of simplicity, and a clearer separation between the particular and the
general propositions of Arithmetic.[77]

Aside from these considerations, however, what is important is that the axioms
for mathematical induction are interdeducible between the systems of all three,
Peirce, Dedekind, and Peano.

1.7. Arithmetical Operations

In order to provide for elementary arithmetic, it is necessary to augment
one's description of the natural numbers with some account of the operations of
addition, multiplication, etc. The modern technique of introducing these operations through recursive definitions probably originated with Herman Grassmann
in 1861.[78] And Peirce's 1881 paper is recognized as one of the earliest examples
of this technique.[79]

Dedekind's definitions of addition and multiplication precisely parallel those
given by Peirce. For example, Peirce gives this definition of addition: "By $x + y$
is meant, in case $x = 1$, the number next greater than y; and in other cases, the
number next greater than $x' + y$, where x' is the number next smaller than x."
(3.262) Dedekind's parallel definition states that the sum $m + n$ is "completely
determined" by the two conditions:

$$m + 1 = m'$$
$$m + n' = (m + n)'$$

where the *successor* of a number n is indicated by n' (§135). It is clear that the
two definitions are equivalent – which is also the case for their analogous definitions of multiplication (3.263 and §147). Given addition and multiplication, the
remaining arithmetical operations of subtraction, division, and exponentiation
are easily defined. It should be noted that while Peirce simply assumes the adequacy of recursive definitions, Dedekind establishes this adequacy as a general
theorem (§126). Within the context of the Peirce-Dedekind-Peano axioms, the
principle of definition by recursion is actually a consequence of the principle

---

[77]See [**109**, p. 248].

[78][**39**]. See also [**123**, p. 147].

[79]Peirce's use of recursive definitions is cited by Fraenkel and Bar-Hillel in [**31**, p. 293],
where it is stated, without direct evidence, that this use was not known to either Dedekind
or Peano. Grassmann's *Lehrbuch* is mentioned in the text of Peano [**53**, p. 103] but not at
all in Peirce or Dedekind. Although Peirce was familiar with other work of Grassmann (e.g.,
see 3.152, 3.242n, 4.668), he seems to have been unacquainted with the *Lehrbuch* in 1881.

of mathematical induction; both principles involve the same underlying idea of minimalization.[80]

This derivation of the principle of definition by recursion, however, requires a formalization capable of expressing the full principle of mathematical induction. For this reason, attempts to axiomatize arithmetic within the constraints of first-order logic, without set variables, have typically introduced the definitions of addition and multiplication as extra axioms. Such systems, referred to today as *first-order Peano arithmetics,* use a schema for mathematical induction, thus restricting it to properties expressible within the formal language.[81] On the other hand, axiom systems like those of Peirce, Dedekind and Peano use the full principle of mathematical induction, applying to all sets, and are generally referred to today as *second-order Peano arithmetics.*[82] Such second-order systems allow the reduction of recursive definitions to explicit definitions, and so are not extended by providing extra axioms for addition and multiplication.[83] They are usually thought to require formalization within logics of at least second-order, i.e., using quantification over sets.

The notion that a second-order logic is required in order to express the full mathematical induction principle is disputed by W. V. O. Quine who argues that this really depends upon whether or not first-order logic can take sets as values of bound variables. For Quine, this is a question of ontology – a question that needs to be answered at the level of set theory before being addressed by logic.[84] Quine describes second-order logic as "set theory in sheep's clothing."

---

[80]That each principle entails a justification of the other is shown in [**10**, pp. 115–123].

[81]For example, the axioms for a first-order Peano arithmetic (beginning with 0 and using $'$ as a successor function) might look like this:

1. $\forall x (0 \neq x')$
2. $\forall x \forall y (x' \neq y' \Rightarrow x = y)$
3. $\forall x (x + 0 = x)$
4. $\forall x \forall y (x + y' = (x + y)')$
5. $\forall x (x \times 0 = 0)$
6. $\forall x \forall y (x \times y' = ((x \times y) + x))$
7. Induction Schema: $(U(0) \,\&\, \forall x(U(x) \Rightarrow U(x'))) \Rightarrow \forall x(U(x))$ where U is a formula in the language that contains 'x', and perhaps other variables, free.

The crucial points to note are that the definitions of addition and multiplication, in 3) – 6), must be given as axioms; and that mathematical induction is expressed as a schema which actually stands for an infinite sequence of (first-order) axioms, one for each formula U in the language.

[82]In 1980, I referred to first and second-order Peano arithmetics as, respectively, Hilbert arithmetics and Peano arithmetics. This usage was recommended in [**31**, pp. 292ff.], but is less common today. –Paul

[83][**10**, p. 127]

[84]Quine, *Set Theory and its Logic* [**97**, pp. 28–34]. Also see the discussion in Beth [**10**, pp. 224–228].

He notes that the stratification of logic developed, historically, as an analogue to Russell's theory of types for set theory, and suggests that there is little to be gained by continuing this confusion of logic and set theory:

> Such assimilation of set theory to logic is seen also in the terminology used by Hilbert and Ackermann and their followers for the fragmentary theories in which the types leave off after finitely many. Such a theory came to be called the predicate calculus (Church: functional calculus) of nth order (not to be confused with order in Russell's sense), where n is how high the types go. Thus the theory of individuals and classes of individuals and relations of individuals was called the second-order predicate calculus, and seen simply as quantification theory with predicate letters admitted to quantifiers. Quantification theory proper came to be called the first-order predicate calculus. It was a regrettable trend. Along with obscuring the important cleavage between logic and "the theory of types" (meaning set theory with types), it fostered an exaggerated if foggy notion of the difference between the theory of types and "set theory" (meaning set theory without types) – as if the one did not involve outright assumption of sets the way the other does. And along with somewhat muffling the existence assumptions of the theory of types, it fostered a notion that quantification theory itself, in its 'F' and 'G', was already a theory about classes or attributes and relations. It slighted the vital contrast between schematic letters and quantifiable variables. The notational style that I am deploring was in essential respects Russell's, of course, before it was Hilbert and Ackermann's. It was associated with failures to discriminate propositional functions as open sentences from propositional functions as attributes.[85]

These distinctions, of course, were not available to Peirce or Dedekind, and in Chapter 2 we will interpret their axiomatizations as being full second-order Peano systems. This accords with their approach to mathematical induction, and reflects how both introduced arithmetical operations.

---

[85]See [**97**, p. 258].

1.8. Categoricity

In the concluding sections of this chapter we will assume the results established in Chapter 2, viz., the equivalence of the axiom systems of Dedekind
and Peirce. We will be concerned, instead, with the general problem of the
adequacy of such axiom systems.

To begin with, notice that the primitive terms of our axiom systems, N,
R, f, and 1, can be given meaning in a variety of ways. An interpretation of
these primitives which satisfies the axioms, i.e., according to which all of the
axioms are true, is called a *model* of that axiom system. In particular, models
of the Peano axioms, hence also of the axioms of Peirce and Dedekind, are often
referred to informally as "progressions."

In previous sections we more or less assumed that our axiom systems were
about numbers. In fact, we evaluated various axioms according to how well
they fit with our intuitive conception of "number." But as Bertrand Russell
points out, any progression at all, whether of numbers, or points on a line, or
instants in time, will provide a model for the Peano axioms.[86] And if there is
one such model, there must be an infinite number of them. From the natural numbers, for example, one can construct progressions of multiples, powers,
primes, coefficients, and so forth, all of which, by suitably interpreting the
notion of "successor," provide models of our axioms. Similarly, any natural
number can be interpreted as the least element, "1," thus providing yet another infinite series of progressions. We will examine, in our next section, how
this overabundance of models can be further narrowed through consideration of
the other logical properties customarily accorded natural numbers. Such cardinal properties serve to help isolate that unique model which most completely
corresponds to our intuitive idea of the number series.

This is not to say, however, that the general description of progressions
is not what we had intended. In fact, Russell remarks that "it is not these
[cardinal] properties that ordinary mathematics employs, and numbers might
be bereft of them without any injury to the truth of Arithmetic and Analysis.
What is relevant to mathematics is solely the fact that finite numbers form a
progression."[87] Constituting a progression satisfactorily accounts for the ordinal properties of the natural numbers, and it is these properties, upon which
most of traditional mathematics is based, that we have primarily had in mind
when speaking of "natural numbers." Now Dedekind defined the finite ordinal
numbers by abstraction, i.e., as what is common to all progressions. Modern

---

[86]See [**109**, p. 126].
[87]See [**109**, p. 241].

theorists, because the size of such an abstractive set appears to risk paradox, prefer to construct an exemplary progression which they then define to be the finite ordinal numbers.[88] But in both cases, the definition of ordinal numbers in terms of progressions seems to imply that all of the latter are alike, so that it is possible, for instance, to choose a *single* progression as exemplar. This is clearly true to the extent that progressions, if they are models of our axioms, must share all the properties ascribed to them by those axioms. One could press the question, though, whether the properties ascribed by our axioms are themselves such as to guarantee similarity of structure, or whether, on the other hand, our axiom systems permit models with dissimilar structures. An important criterion of structural similarity is the notion of *isomorphism*: isomorphic models have identical structures in the sense of being related by a one-to-one correspondence which preserves order.[89] And it is a commonly observed consequence of the minimalization provided by mathematical induction that any two models of our axioms will be isomorphic. In fact, second-order Peano arithmetics in general are often said to be *categoric*, meaning that all of the models of such systems are isomorphic.[90] The value of such categoricity is that it allows the notion of a finite ordinal number system to be unambiguously condensed out of the notion of a progression. For this reason categoricity has traditionally been regarded a measure of the adequacy of any axiom system for arithmetic.[91] The entire situation, however, must be reexamined in the light of Gödel's famous incompleteness theorem of 1931.[92] Gödel demonstrated that any formal system powerful enough to include arithmetic must, if it is consistent, contain undecidable propositions. This means that if our axiom systems are consistent, extending them by the addition of either an undecidable proposition, A, or its negation, $\sim A$, would preserve this consistency. Since consistency is the only condition for having a model, this implies that there must be non-isomorphic models for our original axioms, one corresponding to each of these possible extensions.

It is important not to underestimate the range of Gödel's theorem. The conditions are satisfied by elementary and first-order arithmetic systems.[93] But

---

[88]See [**117**, pp. 128–129].

[89]E.g., two models, $\langle N, R, 1 \rangle$ and $\langle N', R', 1' \rangle$, are isomorphic just in case there is a one-to-one function f from N onto N' such that $f(1) = 1'$ and, for all x and y in N, xRy if and only if $f(x)R'f(y)$. The case is similar when more than one primitive relation is involved.

[90]The notion of categoricity was introduced in 1904 by Oswald Veblen, in [**121**]. Discussions of the concept can be found in [**31**, p. 124] and [**10**, p. 124].

[91]See [**10**, p. 124].

[92]See [**37**]. Gödel's paper actually dealt with ω-consistent systems, and his theorem was strengthened to include consistent systems by J. B. Rosser [**106**].

[93]E.g ., van Heijenoort demonstrates Gödel's theorem with respect to a first-order system S in which mathematical induction is expressed by a schema restricted to *wffs* of the system, [**44**].

since such first-order arithmetics do not contain a full mathematical induction axiom, they had never been considered categorical.[94] In fact, the suspicion that weaker systems might have non-isomorphic models can already be seen in Dedekind's letter to Keferstein. And in 1933, Thoralf Skolem actually constructed a model for first-order arithmetic which is not isomorphic to those usually intended.[95] Gödel's 1931 paper, on the other hand, employed the Peano axioms as formalized within Russell's (non-ramified) type theory.[96] The tremendous impact of Gödel's theorem can only be understood in terms of this apparently paradoxical challenge to the categoricity of such second-order Peano arithmetics.

The general problem in interpreting axiom systems for arithmetic is that the properties or sets for which mathematical induction is defined are restricted to those expressible within the formalization chosen. They do not necessarily include *all* the properties or sets which can be found in the model. This is not difficult to see in the case of weak languages in which the range of expressible properties is rather severely restricted.[97] In the case of second-order Peano arithmetics, though, the adoption of set variables would seem to solve this problem by allowing mathematical induction to be defined for all sets. But any

---

[94]This is pointed out in [**47**, p. 62].

[95]See [**116**].

[96]See [**37**, p. 599]. As Gödel points out in his abstract of [**36**, p. 596], any extension of such a system by the addition of a finite number of axioms still satisfies the conditions of his theorem.

[97]Hao Wang gives the following illustration of how such weak languages can produce non-isomorphic models:

> For example, we can imagine a notation in which every expressible property holds either for only a finite (possibly empty) set of numbers, or for all except a finite set of members. One such notation is obtained if $a_1, a_2, \ldots$ are constant names and every property $F(n)$ can only be a truth-function of finitely many equalities $n = a_i$. For example, no expression in it can represent the set of all odd positive integers.
>
> If we agree to use such a language, we can easily find an unintended model for the whole set of axioms . . . Thus, we can take the domain as consisting of not only the positive integers but in addition the positive and negative fractions of the form $(2b + 1)/2$ (b an integer), 1 as 1, and $a + 1$ as the successor of a. It can be verified that [the non-inductive axioms] are satisfied. Moreover, [the mathematical induction axiom] is satisfied because any property for which "$F(1)$" and "$F(m) \Rightarrow F(m+1)$" hold must hold for all numbers. This is evident with the "true" positive integers. If there were a "positive integer" $a/2$ such that "$F(a/2)$" is not true, "$F((a - 2)/2)$" would also be false, $a/2$ being the "successor" of $(a - 2)/2$ and "$F(m)$" implying "$F(m + 1)$;" similarly, "$F((a - 4)/2)$", "$F((a - 6)/2)$," etc. would all be false. Hence, we would have a property expressible in the given language, such that there is an infinity of numbers which do not possess it, contrary to our assumption. Hence, we get two non-isomorphic models for the weakened Peano axioms[**123**, pp. 155–156].

reference to "all sets" must, if the paradoxes are to be avoided, involve an appeal to some underlying set theory. Thus, second-order Peano arithmetics merely transfer the problem of interpretation onto set theory. In turn, any set theory worth speaking of must itself admit a variety of consistent interpretations, must have, in fact, non-isomorphic models.[98] When set variables are interpreted as ranging over all sets in the model, for instance, we have what is called a *standard model*. When, on the other hand, the range of expressible sets is restricted, e.g., by not consistently interpreting the set-theoretical primitive "$\in$" as designating membership, we have what is referred to as a *non-standard model*.[99] The usual proofs for the categoricity of second-order Peano arithmetics hold only with respect to standard models.[100] The implication of Gödel's theorem, however, is that such axiom systems, if consistent, must also have non-standard models.

The role of non-standard models helps explain away the apparent paradox in Gödel's result. But the challenge to traditional conceptions of categoricity is more than appearance and cannot be explained away. Dedekind's attempt to demonstrate that all of his "simply infinite systems" are similar (§132), for example, is refuted outright. Gödel's theorem, in fact, has the rather startling consequence that our axiom systems, if consistent, must have models of every transfinite cardinality.[101] In general, the notion of "absolute categoricity," along with its implied dichotomy between so-called "one-model theories" and "many model theories," has given way in the post-Gödelean era to various conceptions of relative categoricity. No axiomatizable theory of arithmetic can be absolutely categorical.[102]

In this example, the model $< a_1, a_2, a_3, \ldots ; \ldots, b_3, b_2, b_1; c_1, c_2, c_3, \ldots >$ cannot be eliminated because the intended domain $< a_1, a_2, a_3, \ldots >$ cannot be discriminated by the language. There is no property F that corresponds to either this domain or its complement

[98]The Löwenheim-Skolem theorem actually demonstrates that set theories in which non-denumerable classes can be proven to exist have denumerable models. See [**10**, pp. 488–490, 513–516].

[99]See [**31**, pp. 107, 289].The characterization of "standard" and "non-standard" models is a murky business, the use of these terms varying widely in the literature. In the case of arithmetic systems, a precise formalizable definition, enabling us to strain out the non-standard models, is exactly what Gödel says we cannot have.

[100]Beth points out that isomorphic progressions must be constructed within the *same* (standard) model of the underlying set theory [**10**, p. 515]. The dependence of the meaning of "finite ordinal number" upon the variable meaning of "set" was already remarked by Skolem in 1929. [**47**, p. 61].

[101]This result was established for first-order arithmetic in 1936 by A. Malcev [**62**]. The corollary result for higher-order systems is presented in [**47**, p. 62]

[102]A discussion of various proposals for relative categoricity is contained in [**31**, pp. 291–295] pp. 291–295. Notice that Skolem's arithmetic, since is is defined semantically, has but a single model. By Gödel's theorem, however, such a theory is not axiomatizable. See *Ibid.*, pp. 318, 319.

Thus Gödel's theorem asserts an important limitation upon the adequacy of our axiom systems – to the extent that "adequacy" is measured by the intention to isolate a class of isomorphic models. But this limitation should not be considered catastrophic. We must simply readjust our criteria for axiomatic adequacy to fit with more modest and realistic aspirations. Our axiom systems *are* categoric with respect to standard models – which is not an insignificant result, since most of traditional mathematics can be reconstructed by reference to these same standard models. As Carl Hempel points out, "[Gödel 's theorem] does not ... affect the result outlined above, namely, that it is possible to deduce, from the Peano postulates and the additional definitions of non-primitive terms, all those propositions which constitute the classical theory of arithmetic, algebra, and analysis."[103] From this point of view, the presence of non-standard models gives us more than we really wanted, but does not take away anything we already had. It is still possible to derive arithmetic from our axioms.

1.9. Cardinality

Numbers are used in everyday language in two quite different senses; as *ordinal numbers* when indicating position in a series, and as *cardinal numbers* when used to describe the "size" of a set, "how many" elements it contains. Finite ordinal numbers, subject to the qualifications of the preceding section, can be defined in terms of progressions. In this section we will examine the notion of finite cardinal numbers and, in particular, how these are defined in the essays of Peirce and Dedekind. In doing this we will shed some light on the complex web of relationships between cardinals and ordinals.

The best known, and simplest, definition of cardinal numbers is the abstractive definition first proposed by Gottlob Frege. Frege defined the number [*Anzahl*] of a set X as the set of all those sets equivalent to X.[104] The beauty of

---

[103]See [**46**, p. 375n].

[104]Frege's actual definition runs like this: "the number which applies to the concept F is the extension of the concept 'equinumerous with the concept F'." [**33**, p. 117]. Frege remarks that such numbers are comparable not to Cantor's *Anzahlen*, which are defined ordinally, but to his conception of the "power" [*Mächtigkeit*] of a set:

> "Every aggregate of distinct things can be regarded as a unitary thing in which the things first mentioned are constitutive elements. If we abstract *both* from the nature of the elements and from the order in which they are given, we get the 'cardinal number' or 'power' of the aggregate, a general concept in which the elements, as so-called units, have so grown organically into one another to make a unitary whole that no one of them ranks above the others."[**18**, p. 74]

this definition is that it so neatly captures our ordinary understanding of cardinal numbers. The number 2, for instance, is commonly held to be the number of, e.g., sides on a coin, senators from a state, wheels on a bicycle,. etc.; in short, it is held to be a property of such sets. Frege accepts this ordinary use, merely redefining "property" extensionally, i. e., as the set of such sets. At the same time, Frege clarifies what is meant by "such sets," viz.: all those sets equivalent to a given set having that property – two sets being considered "equivalent" just in case they can be put into one-to-one correspondence.[105] Thus the cardinal number 2, according to Frege, is the set of all those sets which can be put, into one-to-one correspondence with an arbitrary pair. Such definitions, notice, do *not* presuppose the idea of cardinal number in the definiens, since an appropriate n-membered set can always be defined logically – a "pair," for example, containing an element x, a different element y, and nothing else.

Frege's definition makes no reference to ordinals or progressions. In this respect, Peirce's general approach to cardinality, with the exception of his 1881 paper, was often in agreement with Frege. Possibly the earliest anticipation of Frege's abstractive definition was Peirce's 1867 conception of the "numerical rank" of a set:

> Now let the letters, in the particular application of Boole's calculus now supposed, be terms of second intention which relate exclusively to the extension of first intentions. Let the differences of the characters of things and events be disregarded, and let the letters signify only the differences of classes as wider or narrower. In other words, the only logical comprehension which the letters considered as terms will have is the greater or less divisibility of the classes. . . [106]

Peirce's later (1890–1904) conception of what he called the "multitude" of a collection was similarly independent of the notion of ordinality.[107] And not until rather late in his life (after 1904) did Peirce reexamine the general question of

---

[105]Such a comparison of sets via one-to-one correspondences originated with Bernard Bolzano [**12**, p. 98].

[106]"Upon the Logic of Mathematics," 3.43-44. It is instructive to compare Peirce's intuitive notion of abstraction, "let the differences of the characters of things and events be disregarded," with Cantor's, "the elements, as so-called units, have so grown organically into one another to make a unitary whole that no one of them ranks above the others," and Dedekind's, "if . . . we entirely neglect the special character of the elements . . ." (the latter in reference to the abstraction of ordinals. See [**24**, p. 68].)

[107]3.546, 4.175, Ms 27. One must beware terminological confusion here. See [**91**, p. 26n15].

how cardinals are related to ordinals.[108] In 1881, however, Peirce clearly defines
the finite cardinal numbers on the basis of, his system of "ordinary number,"
i.e. in terms of finite ordinals:

> Let such a relative term c, that whatever is a c of anything is
> the only c of that thing, and is a c of that thing only, be called
> a relative of simple correspondence ...
>
> If every object, s, of a class is in any such relation being
> c'd by a number of a semi-infinite discrete simple system, and
> if further every number smaller than a number c of an s is itself
> c of an s, then the numbers c of the s's are said to count them,
> and the system of correspondence is called a count ... If in any
> count there is a maximum counting number the count is said
> to be finite, and that number is called the number of the count.
> (3.280)

Peirce describes the process of counting as the construction of a one-to-one
correspondence (relative of simple correspondence) between a set and an initial
segment of ordinals.[109] The cardinal number of such a set, what Peirce calls the
"number of the count," is given by the maximum ordinal in the initial segment
with which it is corresponded.

Dedekind's 1888 definition of cardinal numbers precisely parallels this definition of Peirce's. Dedekind first describes a one-to-one correspondence (similarity) between a finite set S and Z_n, the set of numbers less than or equal to
n:

> If S is a finite system. then ... there exists one and ... only one
> single number n to which a system Z_n similar (*ähnliche*) to
> the system S corresponds, this number n is called the *number*
> (*Anzahl*) of the elements contained in S (or also the *degree*
> of the system S) and we say S consists of or is a system of
> n elements, or the number n shows how many elements are
> contained in S. If numbers are used to express accurately this

---

[108]See "Ordinals," 4.331–4.334 dated Nov. 15–16, 1904 by Max Fisch (private com-
munication), 4.335–4.340 dated c. 1905; and "Cardinal and Ordinal Numbers," 4.657–4.681
from c. 1909, the latter containing a "scholastic disputation" on the topic. Part of the reason
for Peirce's long neglect of ordinal theory is that he was not as quick to grasp, as in the
case of "multitudes," the method of extending this theory to the transfinite. Peirce himself
recognized this, as can be seen from his draft letter to Cantor of Dec. 21, 1900, where he
begins by apologizing "because I could not catch the idea of your ordinal numbers," Ms 173.

[109]Peirce does not treat the question of alternative models – we assume that his "ordinary
numbers" constitute an exemplary progression.

determinate property of finite systems they are called cardinal numbers. ...We also say that these elements are *counted and set in order* by f ...[110]

The cardinal number of S is simply the ordinal n from the initial segment Z_n. Notice that Dedekind, like Peirce, intuitively conceives cardinals to be derived from ordinals by analogy to the process of "counting."

Bertrand Russell, in *Principles of Mathematics*, objects to this method of introducing cardinals:

> Dedekind remarks in his preface that many will not recognize their old friends the natural numbers in the shadowy shapes which he introduces to them. In this, it seems to me, the supposed persons are in the right – in other words, I am among them. What Dedekind presents to us is not the numbers, but any progression: what he says is true of all progressions alike, and his demonstrations nowhere – not even where he comes to cardinals – involve any property distinguishing numbers from other progressions. No evidence is brought forward to show that numbers are prior to other progressions. We are told, indeed, that they are what all progressions have in common; but no reason is given for thinking that progressions have anything in common beyond the properties assigned in the definition, which do not themselves constitute a new progression. The fact is that all depends upon one-one relations, which Dedekind has been using throughout without perceiving that they alone suffice for the definition of cardinals. The relation of similarity between classes, which he employs consciously, combined with the principle of abstraction, which.he implicitly assumes, suffice for the definition of cardinals; for the definition of ordinals these do not suffice: we require ...the relation of likeness between well-ordered serial relations.[111]

Russell goes on to point out that, in regards to definition, cardinals and ordinals are completely independent, and that the question of which to deal with first "resolves itself into one of convenience and simplicity."[112] Russell himself begins with cardinals because he considers them less complicated to define.

---

[110]Definition §161 in [**23**, pp. 109, 110].

[111]See [**109**, pp. 249, 250].

[112]*Ibid.*, p. 251. Thus Russell says: "I do not hold it an absolute error to begin with order ..."

In reference to Dedekind, many of Russell's objections make sense. Dedekind's attempt to intuitively abstract the ordinals out of "all progressions," for example, leaves particular finite ordinals undefined. It provides what is common to all progressions, but not what is common to all n^{th} elements of progressions.[113] More to the point, Russell perceives the inconsistency in Dedekind's failure to adopt an abstractive approach where it would appear most effective, i.e., in the definition of finite cardinals. And given the abstractive approach, Russell is certainly correct in holding that the notion of equivalence is simpler than that of isomorphism.

Russell's *Principles of Mathematics*, however, underestimated the difficulties that lay in store for Frege's method of definition by abstraction. In the Frege-Russell approach, for instance, the existence of a cardinal number n depends upon the existence of a set of all sets having n members. The existence of this latter is inferred from a general law of comprehension to the effect that, given any property (e.g., having n members), there exists a set whose members are just those entities having that property.[114] But it is precisely this law of comprehension which was shown unsound by Russell's discovery, after most of his *Principles* was already written, of his famous paradox – that no set exists containing just those sets having the property of not being members of themselves.[115] Later attempts by Russell to simultaneously avoid this paradox, by his theory of types, and yet retain an abstractive approach to cardinals clearly forfeited the simplicity of Frege's original definition. In fact, by restricting the range of abstraction to sets of a given type, Russell's later procedure reduplicated the series of cardinal numbers in every type above classes of individuals.[116]

The situation with cardinals is basically similar to that already mentioned with regards to ordinals. Most modern axiomatic set theories tend to avoid

---

[113]This point is obscure in the cited passage, but clearly stated immediately following it, *Ibid.*, p. 250. Also, cf. p. 243.

[114]As Russell puts it. "a class may be defined as all the terms satisfying some propositional function." *Ibid.*, p. 20. Russell immediately qualifies this law: "There is, however, some limitation required in this statement, though I have not been able to discover precisely what the limitation is." And in Chapter X and Appendix A (devoted to Frege). Russell clearly recognizes the dangers in an unrestricted law of comprehension. See [10, p. 299].

[115]Although *Principles of Mathematics* was published in 1903. Russell points out in his Introduction to the Second Edition that it was largely written in 1900. Russell discovered his paradox in June, 1901, exchanged letters with Frege concerning it in 1902, and incorporated Chapter X. "The Contradiction", into the first edition of *Principles of Mathematics*. See [43, pp. 124–128].

[116]See [125, p. 345] and [97, p. 283].

the abstractive definition of cardinals because the sets required are too big.[117] Instead they construct an exemplary set of each cardinality and then define this set to be the cardinal number of such sets as are equivalent to it. And if one already has at hand the ordinal numbers, as an exemplary progression, the obvious choice for a standard finite set of given cardinality will be the appropriate initial segment of ordinals. For this reason, the procedure of Peirce and Dedekind, defining cardinals in terms of ordinals, tends to prevail in contemporary set theory.[118]

The definition of finite cardinals in terms of finite ordinals is clearly a marriage of convenience. It avoids the unnecessary multiplication of entities and reflects our ordinary language use of ordinal names (numerals) for the purpose of counting. It should not be construed, however, as any indication of the logical priority of ordinals. Although he had in mind abstractive definitions, Russell was correct in principle, viz., that finite cardinals can be defined independently of ordinals. We could choose our exemplary cardinals, for instance, by beginning with an arbitrary unit set and successively adding new elements to form a pair, a triple, and so forth. We could then define the finite ordinals in terms of this series of cardinals. Thus, we would interpret the "least element" to be a unit set, the "successor" of a set X to be the set $X \cup \{\alpha\}$, for some α not contained in X, and a "natural number" to be, in the fashion of Dedekind, the chain of the least element with respect to succession. Unless our universe is finite, this interpretation of primitives will in fact give us a progression.[119] This is what we meant by claiming, in the previous section, that the intended model for our axiom systems could be singled out by reference to its cardinal properties. It is in terms of cardinality that ordinary usage distinguishes the model $< 1, 2, 3, \ldots >$ from other standard models, such as, e.g., $< 2, 4, 6, \ldots >$.

Given the desirability of identifying finite cardinals and ordinals, one might still ask whether any particular identification recommends itself. In standard set theories there is a unique set which is available regardless of how the universe might otherwise be constituted – the empty set $\emptyset$. It makes sense, then, to pick

---

[117]E.g., the class of all unit sets must have as many members as the universal class, which latter cannot exist in Zermelo-Fraenkel set theory, and cannot be a set in von Neumann set theory. See *Ibid.*, p. 284.

[118]See [**1**, pp. 359ff.] and [**41**, p. 99].

[119]All of the Peano axioms will be satisfied with the possible exception of the (third) axiom requiring uniqueness of successors. This axiom will fail if we run out of new elements α. See, e.g., [**97**, pp. 279–280]. The same difficulty holds for the similar progression constructed from abstractive, instead of exemplary, cardinals as in Russell and Whitehead, *Principia Mathematica*, ∗110. (See [**112**, pp. 23–24].) Russell and Whitehead resolved this difficulty, in *Principia*, by appealing to an axiom of infinity, formulating subsequent theorems dependent upon this axiom hypothetically. We will examine this axiom at greater length in the next section.

the unit set $\{\emptyset\}$ as our least element. Similarly, an obvious candidate for an α not contained in X is the set X itself, which suggests we define the successor of X as the set $X \cup \{X\}$. This procedure, first suggested by von Neumann in 1923, gives us the progression:[120]

$$1 = \{\emptyset\}$$
$$2 = \{\emptyset, \{\emptyset\}\}$$
$$3 = \{\emptyset, \{\emptyset\}, \{\emptyset, \{\emptyset\}\}\}$$
$$\dots$$

Each finite number, defined in this way, is both a cardinal and an ordinal. Notice that, since ordinals are themselves construed in the form of initial segments, it is no longer necessary to say, with Peirce and Dedekind, that the cardinal number of a set is the maximum ordinal in the corresponding initial segment. Instead we can say that, for finite sets, the cardinal number *is* the ordinal number. Notice, also, that von Neumann's numbers are conveniently ordered by the primitive membership predicate $\in$ of the underlying set theory.[121]

We have been concerned, thus far, entirely with finite ordinals, i.e., with members of an exemplary progression. It will be instructive to generalize our conception of ordinality by adopting Russell's suggestion, in reference to Dedekind (see above), of defining ordinals without relying upon the notion of a progression. Cantor had originally defined the *ordinal type* of a simply ordered set by appeal to an intuitive act of abstraction.[122] Russell interpreted this abstraction extensionally, after the manner of Frege, so that an ordinal type became a set of isomorphic simply ordered sets. And Russell then defined an *ordinal number* after the fashion of Cantor, i.e., as the ordinal type of a well ordered set.[123] Since well ordered sets are not, like progressions, restricted as

---

[120]See [**122**, pp. 348–354]. This natural construction was impossible for Russell since it violated his type strictures. See [**112**, Chapter XIII].

[121]In fact, von Neumann's numbers constitute a monotone set of initials such as referred to earlier (p. 25) in connection with Dedekind's series of remainders. They have the advantage of not assuming infinite sets for the definition of particular finite numbers (as well as allowing the simple construal of cardinals mentioned in the text).

[122]"If the act of abstraction referred to ... is only performed with respect to the nature of the elements, so that the ordinal rank in which these elements stand to one another is kept in the general concept, the organic whole arising is what I call 'ordinal type,' or in the special case of well-ordered aggregates an 'ordinal number.' This ordinal number is the same thing that I called, in my Grundlagen of 1883, the 'enumeral' [*Anzahl*] of a well-ordered aggregate." (From an 1883 lecture, [**18**, pp. 74, 110–117].)

[123]Russell and Whitehead, *Principia Mathematica*, ∗153, ∗251, described in [**97**, p. 153] and [**112**, p. 93] (where "relation-number" is a generalization of "ordinal type", i.e., not restricted to simple orderings). See also [**18**, p. 152]. It is to this approach that Russell refers in the passage cited above when he says that "for the definition of ordinals ... we require ... the relation of likeness between well- ordered serial relations." [**109**, p. 250].

to length, this procedure provides the requisite generalization of the notion of ordinality.[124] As before, we do best to dispense with abstraction and choose a single representative for each type of well ordered set. The von Neumann numbers obviously recommend themselves for this task, each consisting of a family of sets well ordered by the membership relation.[125] And since every von Neumann number is the set of all its predecessors (plus $\emptyset$), it is easy to see how his construction can be extended beyond the finite. The first transfinite ordinal is merely the set ω of all finite ordinals (plus $\emptyset$), the second is $\omega \cup \{\omega\}$, the third is $w \cup \{\omega\} \cup \{\{\omega\}\}$, and so on until we reach $\omega \cdot 2$ which is the set of all finite ordinals and successors of ω (including, of course, both $\emptyset$ and ω. And so on. Although it requires some proving, it is actually possible to demonstrate that every well ordered set must be isomorphic to some von Neumann ordinal.[126]

This generalization of the ordinals allows us to make an important distinction concerning their relation to cardinals. While it may be desirable to identify the finite cardinals with ordinals, this is *not* the case with regards to transfinite cardinals and ordinals. Thus, the ordinal ω and its successor $\omega \cup \{\omega\}$ although not isomorphic, can easily be put into one-to-one correspondence (e.g., by letting $f(\omega) = \emptyset, f(\emptyset) = \{\emptyset\}$, and mapping subsequent numbers into their immediate successors).

And the same holds for the rest of the ordinals in Cantor's second number class, viz.,ω, $\omega+1$, $\omega+2$, ..., $\omega 2$, ..., $\omega 3$, ..., ω^2 ..., ω^3, ..., ω^ω, ..., ω^{ω^ω}, ...; ϵ_0, $\epsilon_0 + 1$, $\epsilon_0 + 2$, ..., $\epsilon_0 + \omega$, etc.[127] These ordinals are said to be denumerable, and in particular, since they are all equivalent to ω, to be *denumerably infinite*. By virtue of their well ordering the ordinals beyond this class, if there are any, must include a least element. The least *non-denumerable* ordinal, i.e. the first ordinal with a cardinality greater than ω is commonly designated ω_1. It begins a further class of equivalent ordinals, viz., ω_1, $\omega_1 + 1$, $\omega_1 + 2$, ..., $\omega_1 + \omega$, ...etc. Likewise, $\omega_2, \omega_3, \omega_4$, and so on – assuming the appropriate classes are not empty – denote the initial ordinals in number classes of progressively greater and greater cardinality. We obtain, in short, a sequence of transfinite number classes, each containing many equivalent (but not isomorphic) ordinals. And since equivalent sets must be assured a *unique* cardinal number, it is clearly not possible to let every such ordinal be an exemplary cardinal.

---

[124]Our axioms require that progressions be 1) infinite, and 2) minimalized. See the section on mathematical induction, above.

[125]The exact definition – X is a (von Neumann) ordinal just in case every member of X is also a subset of X and X is well ordered by membership – was first given [105]. See [97, p. 157n].

[126]This so-called "enumeration theorem" depends upon the axiom of replacement. see [117, p. 234, Theorem 81].

[127]See [18, pp. 160ff.].

In response to these considerations, contemporary set theories generally define the cardinal number of a set to be the least ordinal with which it is equivalent.[128] The customary series of transfinite cardinals, the alephs, $\aleph_0$, $\aleph_1$, $\aleph_2$, ..., are thus identified with the series of initial ordinals, ω, ω_1, ω_2, ... This method of extending his earlier definition of cardinals was already noted by Peirce in 1908: "the multitude of a collection is not the last ordinal in this or that count of it, but is the earliest ordinal that can count it, the count varying according to the order of counting."[129] Such an approach is appealing because it captures the overall relationship between cardinals and ordinals, as originally set forth, e.g., in this description by Cantor:

> The conception of number which, *in finito*, has only the background of enumeral, splits, in a manner of speaking, when we raise ourselves to the infinite, into the two conceptions of power [Mächtigkeit] ... and enumeral [Anzahl] ...; and, when I again descend to the finite, I see just as clearly and beautifully how these two conceptions again unite to form that of the finite integer.[130]

Notice that the comparative scarcity of transfinite cardinals – one for each number class – can only reinforce the tendency to define cardinals in terms of ordinals. It provides a natural argument for the simplicity and convenience of the Peirce-Dedekind approach as opposed to the Frege-Russell approach. Although it would be a mistake to read this insight back into Peirce's 1881 definition. which was probably motivated by its analogy to ordinary counting, it is clear that Peirce eventually adopted such an argument for the priority of ordinals:

> If therefore, we extend the term "cardinal number" so as to make it apply to infinite collections, a multitude of ordinal numbers will be possible exceeding that of all possible cardinal numbers in any infinitely great ratio you please, without having begun to exhaust the ordinals in the least. The system of ordinals is thus infinitely more rich than the system of cardinals. In fact, those two denumeral series of ordinals which are alone required to count all the cardinals [viz., 1, 2, 3,..., ω, ω_1,

---

[128]This definition does require the axiom of choice in order to prove that every set is equivalent to some ordinal (the "enumeration theorem") and thus that every set has a cardinal number. See [**117**, pp. 225, 241–242] and [**97**, pp. 214–215].

[129]From a letter to Francis C. Russell, September 18, 1908, p. 8 (Ms L387), [**70**, p. 257].

[130]See [**18**, p. 52].

$\omega_2, \ldots]$ seem to the student of this branch of mathematical logic most beggarly.[131]

Thus it seems correct to credit Peirce not only with the first ordinally based definition of finite cardinals, but also with an early recognition of the natural advantages of extending this procedure to the transfinite. In this sense, and despite the intervening period during which he examined "multitudes" without reference to ordinals, Peirce clearly anticipated the dominant modern set theoretical approach to cardinality.

Although it is important for understanding his philosophical thought, further discussion of Peirce's theory of transfinite numbers lies beyond the scope of this section.[132] Our purpose has been to merely to situate Peirce's 1881 definition of cardinals with respect to the subsequent development of set theory. All three axiom systems – by Peirce, Dedekind, and Peano – are satisfied by progressions, any standard progression will serve for finite ordinals, and finite cardinals are conveniently defined by reference to ordinals. There is leeway for choice among exemplary progressions, and we have presented the von Neumann numbers instead of, e.g., the Zermelo numbers. But, as Quine points out, this generality should not be construed as an insufficiency:

> Any objects will serve as numbers so long as the arithmetical operations are defined for them and the laws of arithmetic are preserved. It has sometimes been urged that more is wanted: it is not enough that we account for pure arithmetic, we must also account for the application of number in the measurement of multiplicity. But this position, insofar as it is thought of as contrary to the other, is wrong. We have seen how to define not only the arithmetical operations but also the *Anzahlbegriff*, 'α has x members', without having yet decided what numbers are.[133]

Given any standard model for the axioms, one can define all of the properties, cardinal and ordinal, customarily accorded the natural numbers.

---

[131]4.674 (1909). Peirce considers only those cardinals less than $\aleph_w$ and he often confuses this series with the cardinals obtained by successive applications of Cantor's theorem. See [**91**, p. 26].

[132]Peirce's theory of transfinite numbers is discussed in [**70**, pp. 238–288]. The main issues we have with this work are that: 1) Murphey relies too heavily on a single manuscript, Peirce's 1908 letter to Judge Russell, and otherwise tends to telescope Peirce's work, viewing chronological changes as outright contradictions, and 2) Murphey does not pay enough attention to problems connected with terminology. On this topic see also [**91**, pp. 25–27].

[133]See [**97**, p. 151].

1.10. The Existence of a Model

As is evident from our discussion thus far, there are a number of similarities between the essays of Peirce and Dedekind. Both of their axiom systems, despite the fact that they begin with different primitives, converge conceptually – to such an extent that, in our next chapter, we shall demonstrate their equivalence. Both essays derive the expected arithmetical laws, viz., associativity, commutativity, distributivity, etc., both define what it means for a set to be (both ordinary and Dedekind) infinite; both provide recursive definitions of the arithmetical operations; and both define cardinals on the basis of ordinals. The similarities are striking enough to lend credence to the suggestion that Dedekind's essay was actually influenced by Peirce's 1881 paper.[134]

There is one important respect, however, in which the essays of Peirce and Dedekind are not similar. Although both essays provide an axiom system for the natural numbers, Dedekind goes beyond merely describing these numbers axiomatically and tries to show that there must be a model satisfying his axioms. In contrast to Peirce, Dedekind attempts to demonstrate that there must actually exist a set N, a function f, and an element 1, such that all the requirements for being a "simply infinite system" are met.

The crux of Dedekind's argument is his proof – similar to a proof first proposed by Bolzano in 1851 – of a theorem asserting the existence of a (Dedekind) infinite sets:[135]

> 66. Theorem. There exist infinite sets.
> Proof. My own realm of thoughts [*meine Gedankenwelt*] i.e., the totality S. of all things which can be objects of my thought, is infinite. For if s signifies an element of S, then the thought s', that s can be an object of my thought, is itself an element of S. If we regard this as the image $f(s)$ of the element s, then the function f of S, thus determined, has the property that the image S' is a subset of S; and S' is certainly a proper subset of S, because there are elements in S (e.g., my own self) which are different from each such thought s' and therefore are not contained in S'. Finally, it is clear that if a

---

[134]Peirce later claimed that Dedekind's *Was sind und was sollen die Zahlen?* "contains not a single idea which was not in my paper of 1881, of which an extra copy was sent to him and I do not doubt influenced his work." (Ms 316a–s, 1903) Our own position is merely that, judging from internal evidence, such an historical influence would make sense. Dedekind's essay obviously has independent merits, e.g., its pioneering use of set theory.

[135]See [**12**, pp. 84–90].

and b are different-elements of S, then their images, a' and b',
are also different, that therefore the function f is one-to-one.
Hence S is infinite, which was to be proved.[136]

Dedekind argues that his own realm of thoughts must be infinite since it can be
put into one-to-one correspondence with a proper subset of itself by a function
resembling reflection. The step from the existence of such a set to the existence
of a simply infinite system is then straightforward. Dedekind proves (§72) that
every infinite set must have a subset which is simply infinite, namely, the chain
of the difference between that set and its image. Thus, in the example given, the
chain of "my own self" would satisfy all of Dedekind's axioms for being a simply
infinite system. And because this chain is a subset of *meine Gedankenwelt*, its
existence can be deduced from the existence of the latter.

Dedekind's theorem played an important role in the subsequent development of foundations. But his attempt to prove this theorem had little impact,
and in the twentieth century his theorem has usually been formulated as an
axiom – the so-called "axiom of infinity." It will be helpful to look a bit closer
at the history of this axiom. Along with Frege, Bertrand Russell had originally
thought that the existence of a model for the natural numbers could be established by purely logical means. Thus, Russell argued in 1903 that, since the set
of cardinals from 0 to n must always have a cardinality greater than n, the set
of all finite cardinals (including 0) must be infinite. He said. "that there are
infinite classes is so evident that it will scarcely be denied. Since, however, it is
capable of formal proof, it may be as well to prove it."[137] But by 1919, Russell
was telling a different story:

> The conclusion is, therefore, to adopt a Leibnizian phraseology,
> that some of the possible worlds are finite, some infinite, and
> we have no means of knowing to which of these two kinds our
> actual world belongs.[138]

The intervening discovery of the paradoxes had discredited the law of comprehension upon which Russell's argument was based. Without the confidence
that there is a set corresponding to every propositional function, one can no
longer be confident that the number of cardinals from 0 to n is indeed greater

---

[136]See [**73**, pp. 84–90].This is our own translation, incorporating a more contemporary
terminology and, unlike Beman, not neglecting the important *jedem* from the third sentence.
"von jedem solchen Gedanken *s'* verschieden ... sind."

[137]See [**109**, pp. 357–358]. As well as this logical argument, Russell also mentioned the
Bolzano-Dedekind proof.

[138]See [**112**, p. 141].

than n. As remarked in the previous section, it is necessary to first establish
the uniqueness of immediate successors for cardinals, which, in itself, already
presupposes the existence of an infinite set.[139] Nor could Russell have salvaged
his argument by appealing to the von Neumann formulation of successors, since
such a formulation would have violated the type strictures by which he intended
to avoid the paradoxes.[140]

For several years following his discovery of the paradoxes, Russell continued
to hope that a logical proof for the existence of an infinite set could be found.
But in his 1908 paper, "Mathematical Logic as Based on the Theory of Types,"
Russell finally conceded the futility of this endeavor:

> ... suppose there were only n individuals altogether in the uni-
> verse, where n is finite. There would then be 2^n classes of
> individuals, and 2^{2^n} classes of classes of individuals, and so
> on. Thus the cardinal number of terms in each type would
> be finite; and, though these numbers would grow beyond any
> assigned finite number, there would be no way of adding them
> so as to get an infinite number, Hence we need an axiom, so
> it would seem, to the effect that no finite class of individuals
> contains all individuals; but if anyone chooses to assume that
> the total number of individuals in the universe is (say) 10,367.
> there seems no a priori way of refuting his opinion.[141]

Although *Principia Mathematica* was rewritten so as to make those theorems
dependent upon the axiom of infinity explicitly conditional, Russell and Whitehead remained apologetic about the need for this axiom. As a logicist, Russell
was particularly embarrassed about the need to assume not just an infinite
set, but an infinite set of *individuals*, for, as he remarked in 1911, "individual
signifies being of the actual world, as opposed to the beings of logic."[142]

This embarrassment, however, can be attributed to a certain logical pecu-
liarity in the theory of types. Insofar as it regards individuals as constituting
the lowest type, the theory of types requires an infinite number of individuals
in order to get infinite sets of any type (as is implied in the above passage by
Russell). So a model for the natural numbers constructed within type theory,

---

[139]On the importance of the law of comprehension in Cantor, Frege, and Russell,[**11**,
pp. 31–38] and [**10**, pp. 353–360, 366–367, 465ff.].

[140]See [**112**, p. 134].

[141]See [**110**, pp. 150–182, 179]. Russell apparently changed his mind on this point in
1907. See [**40**, pp. 24, 102–107].

[142]See [**111**, pp. 162–174, p 163].

whether consisting of individuals or sets, requires the existence of individuals. On this point the theory of types differs from axiomatic approaches, e.g., Zermelo-Fraenkel set theory, in which the axiom of infinity has no bearing at all upon the question of individuals. Such axiomatic set theories, in fact, often formulate the axiom of infinity simply as the assertion that the set of von Neumann natural numbers exists – which could clearly be the case in a universe without individuals.[143] Thus, to the extent that Russell was worried about the apparent empirical content of the axiom, his problem was peculiar to the theory of types.

On the other hand, there can be little doubt but that the axiom of infinity is extra-logical in a broader sense, and hence that its necessity signified a shortcoming in the logicist program as originally conceived by Frege and Russell. It is important to realize, in this connection, the implications of the axiom for higher set theory. The assumption of an infinite number of individuals in the theory of types, for instance, also guarantees the existence of a hierarchy of transfinite sets, with cardinality $\aleph_0$, $2^{\aleph_0}$, $2^{2^{\aleph_0}}$ and so on, as one considers progressively higher types. (And the same result is obtained in Zermelo-Fraenkel set theory, by virtue of the axiom for power sets.) Thus the axiom of infinity magnifies the existence commitments of type theory in two directions: with respect to individuals and with respect to non-denumerable sets. In view of this rather unexpected power, then, it is hardly surprising that Whitehead and Russell felt apologetic about adopting the axiom, even conditionally. As Russell pointed out in 1911, "the axiom of infinity suffices on its own to prove the majority of existence-theorems that we need in mathematics."[144]

In replacing his "proof" with an "axiom," Russell was part of a growing historical consensus which extended as well to the Bolzano-Dedekind proof. The paradoxes, by discrediting the law of comprehension, had also undermined the intuitive set theories of Cantor and Dedekind. For a period, Dedekind even withdrew *Was sind was sollen die Zahlen?* from publication.[145] And his argument for the existence of an infinite set was specifically associated with the paradoxes, as can be seen from this passage written by David Hilbert in 1904:

> R. Dedekind clearly recognized the mathematical difficulties encountered when a foundation is sought for the notion of number; for the first time he offered a construction of the theory of integers, and in fact an extremely sagacious one. However,

---

[143][**117**, pp. 20, 138].

[144]See [**111**, p. 166].

[145]See [**31**, p. 3], even though the most important contributions of his essay were not implicated in the paradoxes at all.

I would call his method transcendental insofar as in proving
the existence of the infinite he follows a method that, though
its fundamental idea is used in a similar way by philosophers,
I cannot recognize as practicable or secure because it employs
the notion of the totality of all objects, which involves an unavoidable contradiction.[146]

A similar criticism of Dedekind's proof can be found in Zermelo's 1908
axiomatization of set theory. Zermelo introduced the axiom of infinity and
then remarked in a footnote that,

The "proof" that Dedekind ... attempts to give of this principle
cannot be satisfactory, since it takes its departure from "the set
of everything thinkable," whereas from our point of view the
domain V itself, according to no. 10 [the axiom of separation],
does *not* form a set.[147]

Hilbert and Zermelo both interpreted Dedekind's proof as positing the existence
of a universal set – a set apparently implicated in the paradoxes and prohibited
not only by type theory but also by the axiomatic approaches of Zermelo and
von Neumann. And to a large extent this criticism has become the conventional
wisdom regarding Dedekind's proof. It is repeated in Emmy Noether's com-
mentary on Dedekind, and is referred to in such standard contemporary works
as Fraenkel and Bar-Hillel's *Foundations of Set Theory*.[148]

But this criticism is mistaken on several counts. To begin with, it is not
at all clear that Dedekind's proof actually involves a universal set. My own
realm of thoughts – the *Gedankenwelt* – is arguably not such a set, since it
excludes precisely those elements which are not, or cannot be, objects of my
thought.[149] Furthermore, even if the *Gedankenwelt* did comprise a universal set,
its existence would not necessarily result in an "unavoidable contradiction." It
is true in Zermelo's set theory that a universal set is rendered impossible by the
axiom of separation. Likewise, Russell's theory of types prevents any set from

---

[146]See [**49**, pp. 130–131].

[147]See [**127**, p. 204n].Van Heijenoort credits this paper with introducing the axiom
(p. 199). even though it clearly appears in Cesare Burali-Forti, "A Question on Transfinite
Numbers," [**16**], earlier in his own volume (p. 109n).

[148]Noether refers to the "widerspruchsvollen Begriff der 'Menge alies Denkbaren"' in
Dedekind's proof[**73**, vol. 3, p.391]. See also [**31**, pp. 81–82].

[149]The determinate, non-universal, nature of the *Gedankenwelt* is a central point in
Royce's defense of Dedekind's proof. See [**107**, pp. 563–564] and [**96**]. On the basis of this
passage, we would also disagree with Kuklick's characterization of Royce's Absolute as "the
class of all classes." See [**57**, p. 377].

crossing type boundaries and including everything. But these are not the only methods of avoiding the paradoxes, and it would be wrong to conclude that because they prohibit a universal set every response to the paradoxes must do so, As a matter of fact, there are set theories, e.g., Quine's "New Foundations," which circumvent the paradoxes without ruling out a universal set.[150] In short, any argument against Dedekind's proof on such grounds ought to completely exhibit the alleged contradiction.

In this sense, Dedekind's proof has suffered from a sort of guilt by association – association with Russell's unsuccessful "proof" as well as with Dedekind's own discredited set theory. Furthermore, these associations have tended to obscure an important point, namely, that in crucial respects Dedekind's proof is not mathematical at all. Unlike Russell's infinite set of cardinals, for instance, the existence of the *Gedankenwelt* is not derived from the law of comprehension but is held to be self-evident much in the manner of Descartes' *cogito*. Similarly, the argument that this set is infinite is based upon a particular understanding of the nature of reflection, and is mathematical only in its formulation of what the conditions are for being an infinite set. It was in recognition of these extra-mathematical dimensions that Hilbert referred to Dedekind's method as "transcendental." And Suppes has rather aptly characterized the proof: "a beautiful combination of mathematical reasoning and vague epistemology."[151] In any case, it is apparent that, aside from the issue of possible involvement with the paradoxes, Dedekind's proof should be considered the business more of the philosopher than the mathematician. As Russell was aware, specific epistemological arguments must be adduced in order to effectively criticize or support it.[152]

Dedekind himself did not provide any such arguments. Presumably he would have defended his understanding of the *Gedankenwelt* by simply appealing to the evidence of introspection. But a more sophisticated philosophical context appeared at the turn of the century in the writing of Josiah Royce. In his Supplementary Essay to *The World and the Individual*, Royce adopted Dedekind's proof as the centerpiece of his argument against F. H. Bradley.[153]

---

[150]See [96].

[151]See [117, p. 138].

[152]E.g., [112, p. 139]. See below for further discussion of Russell's criticism.

[153]Bradley had argued that the unity which appearance finds in the Absolute must remain inaccessible to our understanding. Any attempt to understand this unity through relational thought, according to Bradley, is futile, since every relation defined requires new relations to make it comprehensible. Royce sought to counter this argument by showing that the "endless fissions" of relational thought could be unified by reference to a single purpose, an "internal meaning" which develops itself more or less dialectically. And Royce employed Dedekind's *Gedankenwelt* as the prime example of how we can understand infinite processes as the expression of a single unified whole. See [107, pp. 473–512].

Although a full treatment of Royce's essay would take us too far afield, we will illustrate the general tenor of his position by showing how he would respond to some typical criticisms of Dedekind's proof.

Before we continue, though, it needs to be pointed out that Royce makes an adjustment in Dedekind's proof. He objects to Dedekind's choice of which element to leave out of the image of the *Gedankenwelt*. Royce does not think that "my own self" [mein eigenes Ich] can be this element because, for Royce, the self *is* the *Gedankenwelt*. On his view, the self is the entire original set – an opinion which Royce registers by inserting an editorial question mark into his translation of Dedekind.[154] Royce says, for instance, that "the whole determined *Gedankenwelt*, if present at once, would be a self, completely reflective regarding the fact that all of these thoughts were its own thoughts."[155] And in line with this view, Royce remarks that the number system must be "a purely abstract image, a bare, dried skeleton, as it were, of the relational system that must characterize an ideally completed Self".[156] What bothers Royce is that Dedekind initially describes a rich, apparently inexhaustible world of his thoughts, and then adds, parenthetically, this incongruous reference to his own self as a dispensable element of this world. Dedekind seems to have in mind an empty Cartesian sort of ego which just sits there and exists. But for Royce such a notion is nonsensical, and the ego or self must be considered infinite, akin, in this respect, to the Absolute: "Unless the Absolute is a Self, and that concretely and explicitly, it is no Absolute at all. And unless it exhausts an infinity, in its presentations, it cannot be a Self."[157] Since it is obviously essential to Dedekind's proof that *some* thought not be contained in the image of the *Gedankenwelt*, it is important that Royce supply an alternative to Dedekind's proposal. Royce does not view this as a serious problem, and suggests that any "primal thought" whose object is not itself an element of the *Gedankenwelt* will suffice. The example he gives is the thought, "Today is Tuesday."[158] In general, Royce appears to be correct in holding that such alternatives are readily available: it would be altogether too optimistic to maintain that all thought is reflective.

A typical argument against Dedekind's proof, in this case by Bertrand Helm, focuses upon the legitimacy of the assumed one-to-one correspondence between the *Gedankenwelt* and its image. Helm claims that the distinction between a thought, s, and the thought of that thought, $f(s)$ or s', must inevitably break down:

---

[154]I.e., "(for example, my own Ego)(?)," [**107**, pp. 511].

[155]*Ibid.*, p. 534

[156]*Ibid.*, p. 526

[157]*Ibid.*, p. 580

[158]See [**107**, p. 533].

Consider. Let "man is a featherless biped" be an element in my thought world. Let me have another thought about that thought: "I am thinking that (or, I have the thought that) man is a featherless biped." As I think over this thought, I find that its content is identical with the first thought, i.e. that man is a featherless biped. With a little effort, and in an attempt to discover an s' here, I might then add that I have the thought that man is such a thing. Here, s', the element that was not before in my thought world in connection with featherless bipeds and men, refers to the addition "I have the thought that ..." But this s' has application, not only to s. here, but to many other elements in my thought world that I might catch myself rethinking and reflecting that I have thought them before. In all such cases, where is the required new thought?[159]

A prima facie case against Helm's argument is not difficult to constructs Thinking that man is a featherless biped and reflecting upon this same thought are different because the circumstances in which the two thoughts would be likely to arise are different. The former might be occasioned by a variety of natural observations while the latter would be more likely to occur in some reflective enterprise such as the study of Aristotle. And if one can understand the difference between these types of occasion, then it would seem that one ought to be able to understand the difference between s and s'.

Helm does not consider this sort of objection, even though it appears to point toward a rather fundamental difficulty in his argument, viz., the barrenness of his conception of reflection. Notice, for instance, Helm's characterization of the elements, s', as those "that I might catch myself re-thinking, and then reflecting that I have thought them before." His model of reflection clearly amounts to little more than an awareness of temporal repetition. Royce, who was especially concerned that the self *not* be construed in this fashion, consistently argues against such models, and for a richer, more philosophical conception of reflection. Thus, in his Supplementary Essay, Royce remarks that:

> ... reflective selfhood, taken merely as the abstract series, I know, and I know that I know, etc., appears to be a vain repetition of the same over and over. But this it appears merely if you neglect the concrete content which every new reflection, when taken in synthesis with previous reflections, inevitably

---

[159]See [**45**, p. 234]. Helm points out that this argument is analogous to that of Bertrand Russell, cf. below.

implies in the case of every living subject-matter. A life that knows not itself differs from the same life conscious of itself, by lacking precisely the feature that distinguishes rational morality alike from innocence and from brutish naivete. A knowledge that is self-possessed differs from an unreflective type of consciousness by having all the marks that separate insight from blind faith.[160]

For Royce, a thought of a thought is more than just an awareness of redundancy; according to him, it actually introduces new content into our thought. And this view of reflection further reinforces our prima facie argument. For instance, when one reflects upon the thought that man is a featherless biped, one is naturally led to compare this thought with other definitions, and by this path to arrive at such notions as genus and specific difference, accidental and essential predication, etc. These notions are implied by the reflection, but not by the original unreflective thought. Royce would hold that something of this sort is generally the case, that reflection always contributes significantly to our understanding. In this sense, Royce's position merely elaborates a fundamental conviction of philosophers, viz., that the examined life is different from that lived unreflectively. It is remarkable that this conviction can be used to support an axiom of set theory, that it has any repercussions at all for such an abstract and technical discipline.

In contrast to Helm, whose arguments were actually directed against Royce rather than Dedekind, and so ought to have addressed the specific counterarguments contained in the Supplementary Essay, Bertrand Russell was apparently unfamiliar with Royce's work. Russell's criticism of Dedekind's proof begins like this:

> We are then to suppose that, starting (say) with Socrates, there is the idea of Socrates, and so on ad info Now it is plain that this is not the case in the sense that all these ideas have actual empirical existence in people's minds. Beyond the third or fourth stage they become mythical. If the argument is to be upheld, the "ideas" intended must be Platonic ideas laid up in heaven, for certainly they are not on earth. But then it at once becomes doubtful whether there are such ideas. If we are to know that there are, it must be on the basis of some logical theory, proving that it is necessary to a thing that there should

---

[160]See [**107**, p. 587].

be an idea of it. We certainly cannot obtain this result empirically, or apply it, as Dedekind does, to "meine Gedankenwelt" – the world of my thoughts.[161]

This passage is so unabashedly anti-idealistic that it scarcely comes into contact with Royce's position at all. Royce, for example, would find the empirical touchstone of such a series of ideas in our own intention to reflect. Through this intention the intellect seizes upon itself and develops its internal meaning as an infinite series. This process is empirical, but in the way meanings and purposes are empirical instead of in the way Russell would construe the notion. Royce remarks that the process is self-evident and that it "does not depend upon a theory about how thought, as an 'activity', is a possible part of the world at all."[162] Royce is quite emphatic on this point:

> I do not profess now to explain, say from a psychological point of view, the inmost nature of the operation in question, nor yet to find self-evident, in this place, the metaphysics of the time process. Mysteries still surround us; but we see what we see. And my point is that while we do not see all of what thought is, nor yet how it is able to weave its material into harmony with its purposes, nor yet what Time is, we do see that we think, and that this thought has, as it proceeds, its internal meaning, and that this meaning has, as its necessary and self-evident result, the reinstatement, in a new case, of the type of situation which the operation of thought was intended to explain, or in some other wise to transform. When M is so altered by the operation C as to imply M', M'', and so on, as the endless series of results of the iterative operation of thought, we see not only that this is so, but why this is so. And unless we see this, we see nothing whatever, whether in Appearance or in Reality.

In a similar vein, Royce would hardly consider Russell's reference to "Platonic ideas" to be as pejorative as it was intended; and he would admit, on the basis of a developed philosophical perspective, that things must be known. In short, there is not all that much common philosophical ground between Russell and Royce. Hence the the issue of Dedekind's proof would have to revert, in part, to a discussion of these more general differences.

---

[161]See [**112**, p. 139].
[162]See [**107**, p. 498].

But Russell follows the passage cited above with a less polemical argument which does appear immediately relevant to Royce's position. Like Helm, Russell attacks the possibility of a one-to-one correspondence between ideas and their objects. Unlike Helm, he does not claim that the distinction between the two in fact collapses. Instead he argues that either idea and object are identical, or an idea is a description of its object.[163] In the latter case, though, one-to-one correspondence would become impossible for a new reason. As Russell points out, most objects can be described in a variety of ways, e.g., Socrates as "the master of Plato," "the husband of Xantippe," etc.

This argument is relevant to the extent that it implies that the *Gedanken-welt* cannot be composed entirely of descriptions. But this does not really exhaust the possibilities; it does not show that there is no satisfactory alternative conception of the relation of object to idea. Furthermore, it is important to recognize that Royce's position does not require a one-to-one correspondence between *every* idea and its object, but only between *reflective* ideas and those thoughts which they reflect. Those original elements of the *Gedankenwelt* which are not themselves images of other elements, i.e., what Royce calls "primal thoughts," may be related to their objects in any way you please without affecting Dedekind's proof. They may not even have objects – so much is first order epistemology undetermined here. Russell completely misses this point, simply assuming that reflection will conform to our model for knowledge of the external world.

There is obviously much more to be said regarding Royce's defense of Dedekind. Although we are sympathetic with Royce's position, our purpose has been merely to suggest that the soundness of Dedekind's proof is still an open philosophical issue. Certainly the arguments of Helm and Russell, while not as misdirected as the criticisms of Hilbert and Zermelo, are not conclusive.

Dedekind's theorem itself – the axiom of infinity – remains a controversial topic in the philosophy of mathematics. It provides the key to Cantor's theory of transfinite numbers – and Hilbert spoke for many mathematicians when he asserted: "No one shall drive us out of the paradise which Cantor has created for us."[164] On the other hand, many intuitionists reject the axiom of infinity, and would agree more with Hermann Weyl's assessment of Cantorism:

> We must learn a new modesty. We have stormed the heavens,
> but succeeded only in building fog upon fog, a mist which will

---

[163]Russell actually considers a third, "psychological," relation of idea to object which we do not pursue since Royce would obviously reject it. [**112**, p. 140].

[164]See [**50**, p. 141].

not support anybody who earnestly desires to stand upon it. What is valid seems so insignificant that it may be seriously doubted whether analysis is at all possible.[165]

Dedekind's theorem also has implications which extend well beyond mathematics. The question of the existence of an infinite set probably first arose in conjunction with Aristotle's distinction between the potential and actual infinite. This distinction was crucial to many traditional philosophical problems, ranging from Zeno's paradoxes to the medieval discussion of the eternity of the world. And in modern philosophy, the dispute between Royce and Bradley is just one example of the continuing relevance of this question. Other examples could easily be given, culled not only from philosophy but from disciplines as diverse as theology and Physics.[166]

In view of these rather far-reaching implications, it is only natural to be curious about Peirce's attitude toward Dedekind's theorem. Although he does not discuss the question in "On the Logic of Number," it is not difficult to reconstruct his position from other writings. We will conclude this section with a brief sketch of Peirce's general approach to the question of the existence of an infinite set.

To begin with, Peirce avoids this question in his 1881 paper quite deliberately, because he does not consider the existence or non-existence of a model for his axioms to even be a mathematical question. According to Peirce, mathematics is purely "hypothetical," it is concerned entirely with what would necessarily be the case given certain conditions, not with whether such conditions ever actually obtain.[167] In this sense, Peirce would agree not only with our characterization of Dedekind's proof as philosophical, but also with the historical reformulation of Dedekind's *theorem* – by Russell, Zermelo, et. al. – as an *axiom*. We will return to Peirce's overall conception of mathematics again in Chapter 3.

Moreover, *qua* philosopher, Peirce would definitely have reservations about how the axiom of infinity is usually phrased. Existence, for Peirce, comes under the category of secondness. Whatever exists does so arbitrarily, by virtue of brute reactions. But an infinite set, Peirce argues, "contains whatever there can be of some general description, and thus is *by law* and not by blind secondness alone." For this reason, Peirce states that "objects that exist by secondness are

---

[165]Cited in [**20**, p. 229].

[166]Cantor was himself aware of the theological implications of his work. See [**18**, p. 55] and [**20**, Chapter 7]. On the relation between mathematical realism and physics, see, e.g., Hilary Putnam, "What is Mathematical Truth?" [**94**, pp. 60–87].

[167]See 3.560 and 4.240.

necessarily finite in multitude." And, despite the complexity of his conception of the relation between set theory and the categories, Peirce would generally regard the locution "an infinite set exists" with suspicion.[168]

But one could easily rephrase the question in terms more acceptable to Peirce yet still capturing the underlying mathematical issue. Are there infinite sets? To this question, put without reference to the categories, Peirce's answer would be an unequivocal 'yes'. Although he would find it difficult to see how this answer could even be doubted, Peirce would probably accept Dedekind's proof as one possible argument in support of the obvious. As a matter of fact, in 1900 Peirce wrote a review of *The World and the Individual* in which the Supplementary Essay, Royce's defense of Dedekind's proof, was particularly singled out for praise.[169] And in a later review Peirce wrote:

> Mathematicians, perhaps, still linger on the stage, who, in their best days, used to be quite positive that one cannot reason mathematically about infinity, and used to feel, like the old lady about her total depravity, that, this cherished inability being taken away, the bottom would fall out of the calculus. Such notions are obsolete. Various degrees of infinity are today conceived with perfect definiteness; and the utter misapprehension of the metaphysicians about it, above all of Hegel, glares. As a first serious attempt to apply to philosophical subjects the exactitude of thought that reigns in the mathematical sciences, ... Royce's *The World and the Individual* will stand a prominent milestone upon the highway of philosophy.[170]

---

[168]Peirce sometimes describes the transition between secondness and thirdness as being gradual. "As the collection enlarges and the individual distinctions are little by little merged, it also passes out of the realm of brute force into the realm of ideas which is governed by rules." 4.178 (1897), also cf. 4.198, 4.211. Elsewhere Peirce speaks as if every collection were second intentional, an *ens rationis*, see Ms 27, 4.647-650. Peirce's mature conception of set theory would require consideration of what he calls "degenerate cases" of the categories, see 1.521ff.

[169]In [**82**]; reprinted at 8.100–8.107. See especially 8.109, from an unpublished draft for this review.

[170]In [**83**]; reprinted at 8.117, 8.120, 8.126–8.130. For similar remarks see Ms 316a–s. and 6.113–6.114.

CHAPTER 2

Equivalence of the Axioms of Peirce and Dedekind

2.1. Formalization

The formalization and proofs below are based upon an underlying logic and a theory of sets. For the former, we will use a standard predicate calculus with identity; individual variables x, y, z ...; individual parameters a, b, c, ... u, v, w; set variables A, B, C, ...; and a distinguished membership predicate $\in$. In order to describe the mathematical induction axioms of Peirce and Dedekind, we will need to speak of arbitrary sets of natural numbers – which requires quantifying over the set variables.[1]

We will also need some kind of set theory. We will need to allow sets as large as the power-set of the natural numbers. This is not exceptional by set theory standards, and it would not be difficult to handle these sets within any sufficiently strong axiomatic theory, such as Zermelo-Fraenkel set theory or von Neumann-Bernays set theory.[2] But we will adapt, instead, the informal set theory in *Was sind und was sollen die Zahlen?*, modifying it to keep it consistent with the lower reaches of ZF set theory. Since our concern is with the equivalence of two axiomatic structures, existence requirements will often be satisfied by hypothesis.[3]

Definitions will be numbered by the arabic numbers, 1–28, and theorems by the labels, T1–T11. A definition or theorem which corresponds to a definition or theorem from *Was sind und was sollen die Zahlen?* will have the reference next to the right margin (e.g., §21). Peirce's axioms will be labeled P0–P6, and Dedekind's axioms D0–D5. The steps of a proof will be numbered by the small roman numerals, i., ii., iii., except that the basis and inductive steps of a

---

[1] In this sense, one might understand this logic as being second-order. But see the discussion of Quine above, pp. 38–40.

[2] See [**117**] and [**122**].

[3] The existence of a model for the natural numbers is discussed in Section 1.10, above.

mathematical induction will be indicated by (i) and (ii). Each step of a proof will be justified along the right margin by citing previous steps, definitions, theorems, or axioms. When a step reflects an assumption, it will be marked "H" for hypothesis.

2.1.1. Definitions. Definitions 1–10 present the elementary set theory needed for describing relations and functions. Set membership, $x \in A$, is primitive, and sets are constructed either by listing their elements or by abstractive definition. We will interpret the latter according to Patrick Suppes' schema:[4]

$$A = \{x \mid \phi(x)\} \Leftrightarrow \forall x(x \in A \Leftrightarrow \phi(x)) \wedge A \text{ is a set}$$
$$\vee \ (A = \emptyset \wedge \nexists B \forall x(x \in B \Leftrightarrow \phi(x)))$$

An individual, u, differs from the unit set, $\{u\}$, containing that individual.[5]

1. $A \subseteq B = \forall x(x \in A \Rightarrow x \in B)$ §3

2. $A = B = A \subseteq B \wedge B \subseteq A$ §3

3. $A = \emptyset \Leftrightarrow \forall x(x \notin A)$

4. $\cap A = \{x \mid \forall B(B \in A \Rightarrow x \in B)\}$

In order to show that an element x is in $\cap A$, however, it is necessary to show that A contains some set as a member.[6] Otherwise $B \in A$ would always be false, and the condition in the abstractive definition would be true for every x. Hence:

$$x \in \cap A \Leftrightarrow \forall B(B \in A \Rightarrow x \in B) \wedge \exists B(B \in A)$$

This has the consequence, for us, that if A has no member sets, $\cap A = \emptyset$, by the schema for abstractive definition. This differs from the approach of Dedekind, for whom in such a case $\cap A = A$.[7]   (§17)

5. $A \cup B = \{x \mid x \in A \vee x \in B\}$ §8

6. $\{x\} = \{y \mid y = x\}$ §2

7. $\{x, y\} = \{x\} \cup \{y\}$

---

[4]See [**117**, p. 34]. The upshot of this schema is that if x does not exist, the abstraction yields the empty set, $\emptyset$. Abstractive definition can also provide a vehicle for constructing sets explicitly, as in definition 6.

[5]Dedekind conflates these – which is why he uses a single symbol for set membership and for the subset relation. See above, p. 12.

[6]See [**117**, p. 46].

[7]See [**24**, p. 49].

8. $(x, y) = \{\{x\}, \{x, y\}\}$

9. R is a binary relation $\Leftrightarrow \forall x(x \in R \Rightarrow \exists y \exists z(x = (y, z)))$

10. $xRy = (x, y) \in R$

Definitions 11–22 provide enough of the theory of relations for the formalization of Peirce's axioms.

11. $A \times B = \{(x, y) \mid x \in A \land y \in B\}$

12. R is a relation in $A \Leftrightarrow R \subseteq A \times A$

13. R is transitive in $A \Leftrightarrow$
    $\forall x \forall y \forall z(x \in A \land y \in A \land xRy \land yRz \Rightarrow xRz)$

14. R is reflexive in $A \Leftrightarrow \forall x(x \in A \Rightarrow xRx)$

15. R is antisymmetric in $A \Leftrightarrow$
    $\forall x \forall y(x \in A \land y \in A \land xRy \land yRx \Rightarrow x = y)$

16. A is connected by $R \Leftrightarrow$
    $\forall x \forall y(x \in A \land y \in A \Rightarrow xRy \lor yRx)$

17. A is partially ordered by $R \Leftrightarrow$
    R is transitive, reflexive, and antisymmetric in A

18. A is simply ordered by $R \Leftrightarrow$
    A is partially ordered and connected by R

19. $S = \{(x, y) \mid xRy \land x \neq y \land \forall z(zRy \land z \neq y \Rightarrow zRx)\}$

When xSy we say that "x is an immediate successor of y."

20. x is a minimum element in $A \Leftrightarrow$
    $\forall x \forall y(x \in A \land y \in A \land xRy \land yRx \Rightarrow x = y)$

21. x is a maximum element in $A \Leftrightarrow$
    $\forall x \forall y(x \in A \land y \in A \land xRy \land yRx \Rightarrow x = y)$

22. Mathematical induction starting with k

    Let M be an arbitrary set.
    If
    (i) $k \in M$
    (ii) $\forall x \forall y(x \in M \land ySx \Rightarrow y \in M)$
    then
    $$\forall x(xRk \Rightarrow x \in M).$$

Définitions 23–28 introduce the theory of functions required for formalizing the axioms of Dedekind. In definitions 25, and 27–28, the context will unambiguously indicate the function being considered.

23. f is a function $\Leftrightarrow f$ is a relation $\wedge$
$\qquad \forall x \forall y \forall z (xfy \wedge xfz \Rightarrow y = z)$ $\qquad\qquad$ §21

24. f is a function on $A \Leftrightarrow f$ is a function $\wedge$
$\qquad \forall x (x \in A \Leftrightarrow \exists y (xfy))$ $\qquad\qquad$ §21

25. $A' = \{ y \mid \exists x (x \in A \wedge xfy) \}$ $\qquad\qquad$ §21

26. f is one-to-one $\Leftrightarrow f$ is a function $\wedge$
$\qquad \forall x \forall y \forall z (yfx \wedge zfx \Rightarrow y = z)$ $\qquad\qquad$ §26

27. A is a chain $\Leftrightarrow A' \subseteq A$ $\qquad\qquad$ §36

28. $A_\circ = \cap \{ B \mid A \subseteq B \wedge B' \subseteq B \}$ $\qquad\qquad$ §44

Since $\{ B \mid A \subseteq B \wedge B' \subseteq B \}$ is defined to contain only sets, the only way that $A_\circ$ can be empty is if either $A = \emptyset$ or $\{ B \mid A \subseteq B \wedge B' \subseteq B \} = \emptyset$. Otherwise A must be a subset of every such B. In the proofs that follow we will speak of $\{ B \mid A \subseteq B \wedge B' \subseteq B \}$ as "the family of chains containing A." Dedekind calls $A_\circ$ "the chain *of* set A," and permits the chains of individuals as well as of sets.[8] Our approach only allows the chains of sets.

These definitions allow the presentation of Peirce's system of "ordinary number" and Dedekind's "simply infinite system" as structures, which we will refer to, respectively, as Peirce and Dedekind natural number systems.

2.1.2. Peirce's Axiom System. A structure $< N, R, f, 1 >$ is a *Peirce natural number system* if and only if N is a set and 1 is an element such that:

P0. R is a relation on N.
P1. N is partially ordered by R.
P2. N is connected by R.
P3. If an element in N is not a minimum element in N, then it is an immediate successor of some element in N.
P4. 1 is a minimum element in N and there is no maximum element in N.
P5. Mathematical induction starting with k holds in N.
P6. $f = \{ (x, y \mid ySx \}$

---

[8]This is a consequence of Dedekind's conflation of an individual with its unit set.

2.1.3. Dedekind's Axiom System. A structure $< N, R, f, 1 >$ is a *Dedekind natural number system* if and only if N is a set and 1 is an element such that:

D0. f is a function on N.

D1. N is a chain.

D2. $1 \in N \land N = \{1\}_\circ$.

D3. $1 \notin N'$

D4. f is one-to-one

D5. $R = \{(x, y) \mid y \in N \land \{x\} \subseteq \{y\}_\circ\}$

Note that axioms P0 and D0 have been formulated so that the fields of R and f will be confined to the set N.[9]

Inspection will confirm that the axioms P1–P5 capture the spirit of Peirce's text. We have stated these independently rather than cumulatively, so P2 reads "N is connected by R," instead of "N is simply ordered by R," and have introduced the name "1" (as Peirce does at 3.261).

With the exception of D2, the axioms D1–D4 differ only notationally from those in *Was sind und was sollen die Zahlen?*. In D2 we have corrected the conflation of the element 1 with the unit set $\{1\}$, and added the explicit provision that 1 be an element of N.[10] The definitional extensions, P6 and D5, are implicit in the respective texts of Peirce and Dedekind. They do not affect the syntactical strength of either system.[11]

---

[9]It is not clear whether Peirce and Dedekind conceived their primitive relations to be confined to N. Peirce says "In a system in which r is transitive, let the q's of anything include that thing itself, and also every r of it which is not r'd by it." This could mean that q is confined to the system – or that q is confined to being reflexive and antisymmetric within the system. Dedekind is similarly ambiguous, see [28, p. 50]. All that is necessary is that the domain of f be restricted to R; Dl would then account for the range.

[10]Dedekind clearly intended 1 to be an element of N, but his axioms do not guarantee this, either in his own presentation or in the modified ZF version given here. The family of chains containing $\{1\}$ is possibly empty, as explained when discussing definition 28. Hence, we have explicitly conjoined $1 \in N$ to his second axiom. This problem shows the need for something like the ZF *axiom of infinity*, on the set theoretical side, to ensure that the family of chains containing $\{1\}$ is not empty – most easily done by positing that the chain N itself exists. See [161, p. 138] and Section 1.10, above.

[11]See the discussion on pp. 20–24.

2.2. Equivalence

We first prove a few theorems similar to theorems from Dedekind. These theorems depend entirely upon the definitions, and so are independent of the axioms of Peirce and Dedekind. Hence, we will be free to use them in both directions in the following proofs. Unless otherwise specified, set variables refer to arbitrary sets. With respect to a function f, the images and chains of any set may be composed, so that $A_{\circ}'$ will indicate the image of the chain of A, while $A'_{\circ}$ will indicate the chain of the image of A.

Theorems T6–T11 assume that the family of chains containing A is not empty.

2.2.1. Theorems.

T1. $A \subseteq A$ §4

T2. $A \subseteq B \wedge B \subseteq C \Rightarrow A \subseteq C$ §4

T3. If f is a function on N, then

$$A \subseteq B \Rightarrow A' \subseteq B'$$ §22

Proof. Assume that $A \subseteq B$.

   i. $\forall y(y \in A' \Rightarrow \exists x(x \in A \wedge xfy))$ 25
   ii. $\forall y(y \in A' \Rightarrow \exists x(x \in B \wedge xfy))$ H., i
   iii. $\forall y(\exists x(x \in B \wedge xfy) \Rightarrow y \in B')$ 25
   iv. $\forall y(y \in A' \Rightarrow y \in B')$ ii, iii

Hence it follows, by 1, that $A' \subseteq B'$.   □

T4. $(A \cup B)' = A' \cup B'$ §23

Proof.

   i. $\forall y(y \in (A \cup B)') \Leftrightarrow \exists x(x \in (A \cup B) \wedge xfy))$ §25
   ii. $\Leftrightarrow \exists x(x \in A \vee x \in B \wedge xfy)$ 5
   iii. $\Leftrightarrow \exists x(x \in A \wedge xfy) \vee \exists x(x \in B \wedge xfy))$ ii
   iv. $\Leftrightarrow (y \in A') \vee (y \in B'))$ iii, 25
   v. $\Leftrightarrow y \in (A' \cup B'))$ iv, 5

Hence by 1 and 2, $(A \cup B)' = A' \cup B'$.   □

T5. If A is an arbitrary set and $\exists B(B \in A)$, then

$$\forall B(B \in A \Rightarrow B' \subseteq B) \Rightarrow (\cap A)' \subseteq \cap A \qquad \S 43$$

Proof. Assume the antecedent, that $\forall B(B \in A \Rightarrow B' \subseteq B)$:

   i. $\forall x(x \in \cap A) \Rightarrow \forall B(B \in A \Rightarrow x \in B)$ 4
  ii. $\forall B(B \in A \Rightarrow \cap A \subseteq B)$ i, 1
 iii. $\forall B(B \in A \Rightarrow (\cap A)' \subseteq B')$ ii, T3
 iv. $\forall B(B \in A \Rightarrow (\cap A)' \subseteq B)$ H., iii, T2
  v. $\forall x(x \in (\cap A)' \Rightarrow \forall B(B \in A \Rightarrow x \in B))$ iv, 1

Since $\exists B(B \in A)$, it follows that the condition in definition 4 is not vacuously true, so:

  vi. $\forall x(\forall B(B \in A \Rightarrow x \in B) \Rightarrow x \in \cap A)$ $4, \exists B(B \in A)$
 vii. $\forall x(x \in (\cap A)' \Rightarrow x \in \cap A)$ v, vi

And it must follow, by 1, that $(\cap A)' \subseteq \cap A$. Hence we conclude that, if every B in A is a chain and there is at least one set B that is a member of A, then the intersection of A must itself be a chain.   □

$$\lll \qquad \ggg$$

The remaining theorems, T6–T11, depend upon the supposition that the family of chains containing A is not empty, i.e., that $\{B \mid A \subseteq B \wedge B' \subseteq B\} \neq \emptyset$.

T6. $A_\circ' \subseteq A_\circ$ $\S 46$

Proof. By supposition the family of chains containing A is not empty, so there must be at least one chain in this family. It follows directly from 28 and T5, that $A_\circ$ is a chain.   □

T7. $A \subseteq B \wedge B' \subseteq B \Rightarrow A_\circ \subseteq B$ $\S 47$

Proof. Again, since the family of chains containing A is supposed to be non-empty, this follows directly from definition 28. If this family were empty, then by the schema for abstraction T7 would actually remain true – since $A_\circ$ would become $\emptyset$.   □

T8. $A \subseteq A_\circ$ $\S 45$

Proof. Since $\{B \mid A \subseteq B \wedge B' \subseteq B\}$ is not empty, this follows directly from 4 and 28. □

T9. $A_\circ' = A'_\circ$ §57

Proof. By supposition there is a set B such that $A \subseteq B$ and $B' \subseteq B$. By T3 it must follow that $A' \subseteq B'$, and by T2 that A' must also be a subset of B. Hence:

i. $A' \subseteq A'_\circ$ T8
ii. $A'_\circ{}' \subseteq A'_\circ$ T6

Let $K = A \cup A'_\circ$. Then by T4, it follows that $K' = (A \cup A'_\circ)' = A' \cup A'_\circ{}'$.

iii. $K' \subseteq A'_\circ$ i, ii, 5
iv. $K' \subseteq K$ iii, 5
v. $A \subseteq K$ 5
vi. $A_\circ \subseteq K$ iv, v, T7
vii. $A_\circ' \subseteq K'$ vi, T3
viii. $A_\circ' \subseteq A'_\circ$ iii, vii, T2
ix. $A' \subseteq A_\circ'$ T8, T3
x. $A_\circ'' \subseteq A_\circ'$ T6, T3
xi. $A'_\circ \subseteq A_\circ'$ ix, x, T8

Hence, by viii and xi, we have $A_\circ' = A'_\circ$. □

T10. $A_\circ = A \cup A'_\circ$ §58

Proof. By the same reasoning as for T9, step vi., $A_\circ \subseteq A \cup A'_\circ$. But if A is a subset of some chain, then by T8, $A \subseteq A_\circ$, and by our preceding result and T6, $A'_\circ \subseteq A_\circ$. So we must also have $A \cup A'_\circ \subseteq A_\circ$. Hence, $A_\circ = A \cup A'_\circ$. □

T11. Theorem of complete induction §59

Let M be an arbitrary set.
If
(i) $A \subseteq M$
(ii) $(A_\circ \cap M)' \subseteq M$
then
$A_\circ \subseteq M$.

Proof. Assume $A \subseteq M$ and $(A_\circ \cap M)' \subseteq M$. Then,

i.	$A_\circ \cap M \subseteq A_\circ$	4
ii.	$(A_\circ \cap M)' \subseteq A_\circ{}'$	i, T3
iii.	$A_\circ{}' \subseteq A_\circ$	T6
iv.	$(A_\circ \cap M)' \subseteq A_\circ$	ii, iii T2
v.	$(A_\circ \cap M)' \subseteq A_\circ \cap M$	ii, iv, 4

Since the family of chains containing A is not empty, we have by T8, $A \subseteq A_\circ$, and by assumption, $A \subseteq M$, so that:

vi.	$A \subseteq A_\circ \cap M$	T8, 4
vii.	$A_\circ \subseteq A_\circ \cap M$	v, vi, T7

Hence, it follows from the definition of intersection that $A_\circ \subseteq M$. $\qquad\square$

2.2.2. Dedekind $\Rightarrow$ Peirce. Suppose we have a structure $< N, R, f, 1 >$ which is a *Dedekind natural number system,* satisfying the axioms D0-D5. We will derive each of Peirce's axioms P0-P6 as theorems, thus proving that the structure $< N, R, f, 1 >$ is also a *Peirce natural number system.* We start with a few lemmas that will be useful in moving from Dedekind's successor function to Peirce's transitive relation. These depend upon the Dedekind axioms and so can only be used in this direction:

L1. $\forall x \forall y (xfy \Rightarrow x \in N \wedge y \in N)$

Proof. Assume for an arbitrary u and v, that ufv. We know, by D0, that $u \in N$, and we can show that $v \in N$ as follows:

i.	ufv	H.
ii.	$u \in N$	i, D0, 24
iii.	$\{u\} \subseteq N$	ii, 1, 6
iv.	$\{u\}' \subseteq N$	iii, T3
v.	$N' \subseteq N$	D1
vi.	$\{u\}' \subseteq N$	iv, v, T2
vii.	$\{u\}' = \{v\}$	i, 6, 25
viii.	$\{v\} \subseteq N$	vi, vii
ix.	$v \in N$	viii, 1, 6

L2. $\forall x (x \in N \Rightarrow \{x\} \not\subseteq \{x\}_\circ{}')$

Proof. Let M be the set of elements in N such that $\{x\} \not\subseteq \{x\}_o{}'$:

$$M = \{x \mid x \in N \wedge \{x\}\} \not\subseteq \{x\}_o{}'$$

We will show that $1 \in M$, and that the image of any number in M is in M. It will then follow by the theorem of complete induction that all elements of N are in M.

(i) $\{1\} \subseteq M$, since by D2 1 is an element of N, and by D3 1 is not an element of N', which is to say that $\{1\} \not\subseteq \{1\}_o{}'$.

(ii) $(\{1\}_o \cap M)' \subseteq M$. To prove the inductive step, we will suppose the contrary and derive a contradiction. So suppose that $\exists x(x \in (\{x\}_o \cap M)' \wedge x \notin M)$. Call such an element b. From 25 it follows that there must be an element, a, such that $a \in (\{1\}_o \cap M)$ and afb. Then it must be the case that:

i. $a \in M$	H., 25, 4
ii. $b \notin M$	H.
iii. afb	H., 25
iv. $a \in N$	iii, L1
v. $b \in N$	iii, L1
vi. $\{b\} \subseteq \{b\}_o{}'$	ii, v
vii. $\{a\}' \subseteq \{a\}'_o{}'$	iii, vi, 6, 25

D4 guarantees that the function f is one-to-one, so we can prove that $\{a\} \subseteq \{a\}'_o$ as follows:

viii. $\forall x(x \in \{a\}' \Rightarrow x \in \{a\}'_o{}')$	vii, 1
ix. $\forall x(\exists y(y \in \{a\} \wedge yfx) \Rightarrow \exists z(z \in \{a\}'_o \wedge zfx))$	viii, 25
x. $\forall x \forall y \forall z(yfx \wedge zfx \Rightarrow y = z)$	D4
xi. $\forall x \forall y(y \in \{a\} \wedge yfx \Rightarrow y \in \{a\}'_o \wedge yfx)$	ix, x
xii. $\forall y(y \in \{a\} \wedge yfb \Rightarrow y \in \{a\}'_o \wedge yfb)$	xi, (b/x)
xiii. $\forall y(y \in \{a\} \Rightarrow yfb)$	iii, 6
xiv. $\forall y(y \in \{a\}y \in \{a\}'_o)$	xii, xiii
xv. $\{a\} \subseteq \{a\}'_o$	xiv, 1

Since a is in the chain N, we can use T9:

xvi. $\{a\}'_o = \{a\}_o{}'$	iv, T9
xvii. $\{a\} \subseteq \{a\}_o{}'$	xv, xvi
xviii. $a \notin M$	xvii

But this contradicts our assumption, so it must be the case that the inductive step is true, and that $(\{1\}_{\circ} \cap M)' \subseteq M$.

It follows, by the theorem of complete induction, T11, that $\{1\}_{\circ} \subseteq M$, and thus, by D2 and 1, that $\forall x(x \in N \Rightarrow \{x\} \not\subseteq \{x\}_{\circ}')$. □

L3. $\forall x \forall y((x \in N \wedge y \in N \wedge \{x\}_{\circ} = \{y\}_{\circ}) \Rightarrow x = y)$

Proof. Let u and v be arbitrary elements in N and assume that $\{u\}_{\circ} = \{v\}_{\circ}$.

i. $u \in N$	H.
ii. $v \in N$	H.
iii. $\{u\}_{\circ} = \{v\}_{\circ}$	H.
iv. $\{u\}_{\circ}' = \{v\}_{\circ}'$	iii, 2, T3

Since u and v are in the chain N, we can use T8, T9 and T10.

v. $\{u\} \subseteq \{u\}_{\circ}$	i, T8
vi. $\{u\} \subseteq \{v\}_{\circ}$	iii, v
vii. $\{u\} \subseteq \{v\} \cup \{v\}_{\circ}'$	vi, T9, T10
viii. $\{u\} \subseteq \{v\} \cup \{u\}_{\circ}'$	vii, iv
ix. $(u \in \{v\}) \vee (u \in \{u\}_{\circ}')$	viii, 5,6
x. $\{u\} \not\subseteq \{u\}_{\circ}'$	i, L2
xi. $u \in \{v\}$	ix, x
xii. $u = v$	xi, 6

L4. $\forall x \forall y(xfy \Rightarrow ySx)$

Proof. For an arbitrary u and v, assume that ufv.

i. ufv	H.
ii. $u \in N \wedge v \in N$	i, L1
iii. $N' \subseteq N$	D1
iv. $\{v\} = \{u\}'$	i, 6, 25
v. $\{v\}_{\circ} = \{u\}'_{\circ}$	iv

Since u and v are in N, and N is a chain, we can use T8 and T9 and T6.

vi. $\{v\} \subseteq \{v\}_{\circ}$	T8
vii. $\{v\} \subseteq \{u\}'_{\circ}$	v, vii

viii.	$\{v\} \subseteq \{u\}_\circ{}'$	vii, T9
ix.	$\{u\}_\circ{}' \subseteq \{u\}_\circ$	T6
x.	$\{v\} \subseteq \{u\}_\circ$	viii, ix, T2
xi.	$\{u\} \not\subseteq \{u\}_\circ{}'$	ii, L2
xii.	$u \neq v$	vii, xi, 6
xiii.	$(\{v\} \subseteq \{u\}_\circ) \wedge (v \neq u)$	x, xii

Thus by D5 we have $vRu \wedge (v \neq u)$. It only remains to show that for an arbitrary w,

$$\{w\} \subseteq \{u\}_\circ \wedge w \neq u \Rightarrow \{w\} \subseteq \{v\}_\circ$$

Assume the antecedent. Since $\{u\}$ is a subset of the chain N, we must have, by T1, that $\{w\} \subseteq \{u\} \cup \{u\}'_\circ$. But if $w \neq u$, this leaves us with $\{w\} \subseteq \{u\}'_\circ$, which by step iii is the same as $\{w\} \subseteq \{v\}_\circ$. Hence, $\forall z(zRu \wedge z \neq u \Rightarrow zRv)$ and in accordance with 19, it must be the case that v is an immediate successor of u. □

We now derive Peirce's axioms P0–P6.

P0. R is a relation in N

Proof. According to D5, $R = \{(x, y) \mid y \in N \wedge \{x\} \subseteq \{y\}_\circ\}$. Let (u, v) be an arbitrary element of R. By definition $v \in N$, so we have only to show that $u \in N$.

i.	$v \in N$	H.
ii.	$\{u\} \subseteq \{v\}_\circ$	H.
iii.	$N' \subseteq N$	D1
iv.	$\{v\} \subseteq N$	i, 1, 6

Since v is in the chain N, we can use T7.

v.	$\{v\}_\circ \subseteq N$	iii, iv, T7
vi.	$\{u\} \subseteq N$	ii, v, T2

And since for arbitrarily chosen (u, v) in R, we have both $u \in N$ and $v \in N$, it follows that $R \subseteq N \times N$ and R must be a relation in N. □

P1. N is partially ordered by R

Proof. We must show that R is transitive, reflexive, and antisymmetric in N.

i. That R is transitive in N follows directly from D5. Assume that, for arbitrary u, v, and w in N, we have both $\{u\} \subseteq \{v\}_o$ and $\{v\} \subseteq \{w\}_o$. Since N is in the family of chains containing u, v, and w, we can use T6 and T7. By T6 $\{w\}_o$ is a chain, and by T7, $\{v\}_o \subseteq \{w\}_o$. But T2 ensures the transitivity of subsets, so we must have $\{u\} \subseteq \{w\}_o$. Hence, $\forall x \forall y \forall z (xRy \wedge yRx \Rightarrow xRz)$ and R is transitive in N.

ii. That R is reflexive in N follows from D1. Let u be an arbitrary element in N. Since D1 asserts that N is a chain, and $\{u\}$ is a subset of this chain, we can infer by T8 that $\{u\} \subseteq \{u\}_o$. Hence, $\forall x(x \in N \Rightarrow xRx)$ so R is reflexive in N.

iii. That R is antisymmetric in N follows from L3. Assume that both $\{u\} \subseteq \{v\}_o$ and $\{v\} \subseteq \{u\}_o$, for arbitrary u and v in N. Since the family of chains containing u and v includes N, by T6 we know that both $\{v\}_o$ and $\{u\}_o$ must be chains. By T7, then, we must have $\{u\}_o \subseteq \{v\}_o \wedge \{v\}_o \subseteq \{u\}_o$, so by 2 and L3 it follows that $u = v$. Hence, $\forall x \forall y (xRy \wedge yRx \Rightarrow x = y)$ and R is antisymmetric in N.

By i, ii, and iii, then, N is partially ordered by R. $\qquad\square$

P2. N is connected by R.

Proof. The proof is by the theorem of complete induction using D1 and D2. Let u be an arbitrary element of N, and let M be the set:

$$M = \{x \mid x \in N \wedge (\{x\} \subseteq \{u\}_o \vee \{u\} \subseteq \{x\}_o)\}.$$

(i) $\{1\} \subseteq M$, since by D2 we have $1 \in N$ and by hypothesis, $\{u\}$ is a subset of $\{1\}_o$.

(ii) $(\{1\} \cap M)' \subseteq M$. Let v be an arbitrary element in $(\{1\} \cap M)'$. We will prove the inductive step by showing that v must be in M, since $v \in N$ and $\{v\} \not\subseteq \{u\}_o \Rightarrow \{u\} \subseteq \{v\}_o$. By 25, we have $\exists x(x \in (\{1\}_o \cap M) \wedge xfv)$. Call this element a.

i.	$a \in M$	H., 25
ii.	afv	H., 25
iii.	$a \in N \wedge v \in N$	ii, L1

Now assume that $\{v\}$ is not a subset of $\{u\}_o$. Since u, v, and a are elements of N, and by D1 N is a chain, we can use T6, T7, T8 and T10.

iv.	$\{v\} \not\subseteq \{u\}_o$	H.
v.	$\{a\}' \not\subseteq \{u\}_o$	ii, iv, 6, 25
vi.	$\{u\}_o{}' \subseteq \{u\}_o$	T6
vii.	$\{a\}' \not\subseteq \{u\}_o{}'$	v, vi, T2
viii.	$\{a\} \not\subseteq \{u\}_o$	vii, T3

ix. $\{u\} \subseteq \{a\}_\circ$ i ,iii, viii

x. $\{u\}_\circ \subseteq \{a\}_\circ$ ix, T6, T7

xi. $\{u\}_\circ \subseteq \{a\} \cup \{a\}'_\circ$ x, T10

xii. $a \notin \{u\}_\circ$ viii, 1, 6

xiii. $\{u\}_\circ \subseteq \{a\}'_\circ$ xi, xii

xiv. $\{u\}_\circ \subseteq \{v\}_\circ$ ii, xiii, 6, 25

xv. $\{u\} \subseteq \{u\}_\circ$ T8

xvi. $\{u\} \subseteq \{v\}_\circ$ xiv, xv, T2

From iv–xvi, and from the fact that $v \in N$, it must be the case that $v \in M$. This proves the inductive step.

Since $\{1\}$ is a subset of the chain N, by the theorem of complete induction, T11, it must follow that $\{1\}_\circ \subseteq M$. Hence every element of N is an element of M, and because u was chosen arbitrarily, by D5 $\forall x \forall y (x \in N \wedge y \in N \Rightarrow xRy \vee yRx)$, and N must be connected by R.   □

P3. If an element in N is not a minimum element in N, then it is an immediate successor of an element in N.

Proof. For an arbitrary u, assume the antecedent, that u is in N and u is not a minimum element in N. We have, then, $u \in N \wedge \exists x (x \in N \wedge x \neq u \wedge uRx)$. Call this latter element a. We show that u must also be in N':

i. $a \in N$ H.

ii. $a \neq u$ H.

iii. $\{u\} \subseteq \{a\}_\circ$ H., D0

iv. $\{a\} \subseteq N \wedge N' \subseteq N$ i, D1

Since a and u are elements of N and N is a chain, we can use T7, T9, and T10.

v. $\{u\} \subseteq \{a\} \cup \{a\}'_\circ$ iii, T10

vi. $\{u\} \subseteq \{a\}'_\circ$ ii ,v, 5

vii. $\{u\} \subseteq \{a\}_\circ{}'$ vi, T9

viii. $\{a\}_\circ \subseteq N$ iv, T7

ix. $\{a\}_\circ{}' \subseteq N'$ viii, T3

x. $\{u\} \subseteq N'$ vii, ix, T2

xi. $u \in N'$ x, 1, 6

But if u is an element of N', then by 25 we know that $\exists x (x \in N \wedge xfu)$. By L4 u will be an immediate successor of this element, which is also in N.   □

P4. 1 is a minimum element in N and there is no maximum element in N.

Proof. The first half of this axiom follows directly from D2 and the proof that R is antisymmetric in N. By D2, $1 \in N$. Now assume that $\exists x(x \in N \wedge x \neq 1 \wedge 1Rx)$. By D2 and D5 we know that $\forall x(x \in N \Rightarrow xRl)$. But by the antisymmetry of R we cannot have, for such an x, $xR1 \wedge 1Rx \wedge x \neq 1$. Hence, $\forall x(x \in N \wedge x \neq 1 \Rightarrow \neg 1Rx)$ and so 1 must be a minimum element in N.

The second half of the axiom follows from D0, Ll, and L4. Let u be an arbitrary element in N. By D0, f is a function on N, so $\exists x(ufx)$. Calling this element a, we know by Ll that $a \in N$. Hence, by L4 we must have aSu, which according to 19 means that $aRu \wedge a \neq u$. Since u is an arbitrary element in N, then

$$\forall x(x \in N \Rightarrow \exists y(y \in N \wedge y \neq x \wedge yRx))$$

which is the same as saying

$$\nexists x(x \in N \wedge \forall y(y \in N \wedge y \neq x \Rightarrow \neg yRx)).$$

So there can be no maximum element in N. $\qquad\square$

P5. Mathematical induction starting with k holds in N.

We must show that for an arbitrary set M, the following holds:

> If
> (i) $k \in M$
> (ii) $\forall x \forall y(x \in M \wedge ySx \Rightarrow y \in M)$
>   then
> $\forall x(xRk \Rightarrow x \in M).$

Proof. Notice, first, that if $k \notin N$, the conclusion is vacuously true, since the proof of P0 guarantees that both x and k will be in N. So let M be an arbitrary set, let k be an arbitrary element of N, and assume that the basis and inductive steps hold.

i. $k \in N$	H.
ii. $k \in M$	H.
iii. $\forall x \forall y(x \in M \wedge ySx \Rightarrow y \in M)$	H.

Since $k \in N$, the family of chains containing k is not empty and we are able to use T11, the theorem of complete induction. From T11, it will follow that $\{k\}_{\circ} \subseteq M$, since:

(i) $\{k\} \subseteq M$. The basis step comes from ii, above.

(ii) $(\{k\}_o \cap M)' \subseteq M$. Let u be an arbitrary element of $(\{k\}_o \cap M)'$
. According to definition 25, $\exists x(x f u \wedge x \in \{k\}_o \cap M)$. Call this
element a. L4 tells us that uSa, and so by 4, $a \in M$. But then
$u \in M$ by iii, and since u was chosen arbitrarily, it must follow that
the inductive step for the theorem of complete induction also holds,
$(\{k\}_o \cap M)' \subseteq M$.

iv. $\{k\}_o \subseteq M$	(i), (ii), T11
v. $\forall x(x \in \{k\}_o \Rightarrow x \in M)$	iv, 1
vi. $\forall x(\{x\} \subseteq \{k\}_o \Rightarrow x \in M)$	v, 1, 6
vii. $\forall x(xRk \Rightarrow x \in M)$	vi, D5

It follows that whenever the basis and induction steps for Peirce's mathematical
induction axiom obtain, so does its conclusion.   □

P6. $f = \{(x, y) \mid ySx\}$.

Proof. We must show that, for an arbitrary (u, v), ufv if and only if vSu.
The first half of this has been established by L4, so we only need to show here
that $vSu \Rightarrow ufv$. Assume that vSu. From definition 19, it follows immediately
that:

$$vRu \wedge v \neq u \wedge \forall z(zRu \wedge z \neq u \Rightarrow zRv).$$

By the proof of P0 we know that R is a relation in N, so by definition,
$R \subseteq N \times N$, and it follows that u and v are elements of N. Since N is a chain,
by D1, we are able to use theorems T6–T11. We know that:

i. $u \in N$	H.
ii. $v \in N$	H.
iii. vRu	H.
iv. $\{v\} \subseteq \{u\}_o$	iii, D5
v. $v \neq u$	H.
vi. $N' \subseteq N$	D1

Now, let w be an arbitrary element in $\{u\}'_o$. We show that w must also be in
$\{v\}_o$.

vii. $\{w\} \subseteq \{u\}'_o$	H.
viii. $\{u\}'_o \subseteq \{u\}_o$	T6, T9
ix. $\{w\} \subseteq \{u\}_o$	vii, viii

   x. $\{u\}_\circ \subseteq N$ i, vi, T7
   xi. $w \in N$ ix, x
   xii. $w \neq u$ vii, T9, L2
   xiii. wRu ix, xi, D5

But by hypothesis, $\forall z(zRu \wedge z \neq u \Rightarrow zRv)$, so it must also be the case that wRv:

   xiv. wRv xii, xiii, H.
   xv. $\{w\} \subseteq \{v\}_\circ$ xiv, D5
   xvi. $w \in \{v\}_\circ$ xii

Since w was chosen arbitrarily, steps vii to xvi show that $\forall x(x \in \{u\}_\circ{}' \Rightarrow x \in \{v\}_\circ)$ and, by definition 1, that $\{u\}'_\circ \subseteq \{v\}_\circ$. But we can also show that $\{v\}_\circ \subseteq \{u\}'_\circ$.

   xvii. $\{v\} \subseteq \{u\} \cup \{u\}'_\circ$ iv, T10
   xviii. $\{v\} \subseteq \{u\}'_\circ$ v, xvii
   xix. $\{v\}_\circ \subseteq \{u\}'_\circ$ xviii, T7

Thus, $\{u\}'_\circ = \{v\}_\circ$. Now, by D0 there must exist some element, a, in N such that ufa. By 25, $\{u\}_\circ = \{a\}$, by extensionality $\{a\}_\circ = \{v\}_\circ$, and by L3, $a = v$. Hence, ufv. $\qquad \square$

These proofs show that any structure $< N, R, f, 1 >$ which is a *Dedekind natural number system* must also be a *Peirce natural number system*.

2.2.3. Peirce $\Rightarrow$ Dedekind. Suppose we have a structure $< N, R, f, 1 >$ which is a *Peirce natural number system*, satisfying the axioms P0–P6. We will derive each of the axioms D0–D5 as theorems, thus showing that the structure $< N, R, f, 1 >$ is also a *Dedekind natural number system*.

D0. f is a function on N

Proof. We must show that f is a function, and that $\forall x(x \in N \Leftrightarrow \exists y(xfy))$.

We first show that f is a function. By P6, f consists solely of ordered pairs, so f is a relation. Now assume that

$$\exists x \exists y \exists z(ySx \wedge zSx \wedge y \neq z).$$

Let a, b, and c be one such triple. Then according to definition 19,

$$bRa \wedge b \neq a \wedge \forall z(zRa \wedge z \neq a \Rightarrow zRb)$$

and

$$cRa \wedge c \neq a \wedge \forall z(zRa \wedge z \neq a \Rightarrow zRc).$$

By instantiating c for the variable z in the former, and b for the variable z in the latter, we obtain both cRb and bRc. But since P0 guarantees that both b and c are elements of N, the antisymmetry clause of Pl implies that $b = c$ which contradicts our hypothesis. So it has to be the case, rather, that

$$\forall x \forall y \forall z(ySx \wedge zSx \Rightarrow y = z).$$

Hence, by P6

$$\forall x \forall y \forall z(xfy \wedge xfz \Rightarrow y = z),$$

and f must be a function.

We now show that $\forall x(x \in N \Leftrightarrow \exists y(xfy))$. The necessity follows directly from P6 and P0. If, for an arbitrary u, there exists an element a such that ufa, then by P6 aSu, by 19 aRu, and by P0, $u \in N$. So we need only prove the sufficiency, that:

$$\forall x(x \in N \Rightarrow \exists y(xfy)),$$

that is, we need to prove that every element in N has an immediate successor. In order to show a contradiction, let u be an arbitrary element in N, and assume that u does not have an immediate successor. We can use P5, Peirce's mathematical induction axiom, to show that every element greater than u also lacks an immediate successor. Let M be the set:

$$M = \{x \mid x \in N \wedge \nexists y(ySx)\}$$

(i) $u \in M$. By hypothesis u is in M.
(ii) $\forall x \forall y(x \in M \wedge ySx \Rightarrow y \in M)$. The inductive step is vacuously true, since if an element is in M, then it has no immediate successor.

It follows by mathematical induction starting with u, that

i. $\forall x(xRu \Rightarrow x \in M)$ (i), (ii), P5

This means that no element greater than u could have an immediate successor. But there must be an element greater than u since $u \in N$ and, according to P4, N has no maximum element. Hence:

$$\exists x(x \in N \wedge x \neq u \wedge xRu).$$

Call such an element a. We will show that a has an immediate predecessor, and that this predecessor must be both an element of M and not an element of M.

ii. $a \in N \wedge a \neq u \wedge aRu$	i, P4, 21
iii. $u \in N \wedge a \neq u \wedge aRu$	ii, H.
iv. $\exists x(x \in N \wedge x \neq a \wedge aRx)$	iii
v. a is not a minimum element of N	iv, 20

By P3, then, a is the immediate successor of some element, b, in N.

vi. $b \in N \wedge aSb$	v, P3
vii. $b \in N \wedge \exists y(ySb)$	vi
viii. $b \notin M$	vii
ix. $\forall z(zRb \wedge z \neq b \Rightarrow zRa)$	vi, 19
x. $uRb \wedge u \neq b \Rightarrow uRa$	ix, u/z
xi. $aru \wedge a \neq u$	ii
xii. $\neg uRa$	xi, P1
xiii. $\neg(uRb \wedge u \neq b)$	x, xii
xiv. bRu	xiii, P1, P2
xv. $b \in M$	xiv, i

But this is a contradiction, so for arbitrary u in N it must be the case, rather, that $\exists y(ySu)$. Thus every element in N has an immediate successor, and by P6:

$$\forall x(x \in N \Rightarrow \exists y(xfy)).$$

According to 24, then, f is a function on N. $\square$

D1. $N' \subseteq N$.

Proof. This follows directly from the fact that R is a relation in N. If u is an arbitrary element of N', then by 25 there must be some element a such that afu. By P6, uSa, and by 19, uRa. But P0 asserts that $R \subseteq N \times N$, and thus u must be an element of N. Hence, $N' \subseteq N$. $\square$

D2. $1 \in N \wedge N = \{1\}_{\circ}$.

Proof. $1 \in N$ follows immediately from P4. The proof that $N = \{1\}_{\circ}$ is derived chiefly from Peirce's mathematical induction axiom and properties of the element 1. Let $k = 1$, and consider the statement of P5:

Let M be an arbitrary set:

If
  (i) $1 \in M$
  (ii) $\forall x \forall y (x \in M \wedge ySx \Rightarrow y \in M)$
  then
    $\forall x (xR1 \Rightarrow x \in M)$.

The inductive step can be reformulated since

$$
\begin{aligned}
\forall x \forall y (x \in M \wedge ySx \Rightarrow y \in M) &\Leftrightarrow \forall y (\exists x (x \in M \wedge ySx) \Rightarrow y \in M) \\
&\Leftrightarrow \forall y (\exists x (x \in M \wedge xfy) \Rightarrow y \in M) \quad \text{P6} \\
&\Leftrightarrow \forall y (y \in M' \Rightarrow y \in M) \quad\quad\quad\quad 25 \\
&\Leftrightarrow M' \subseteq M \quad\quad\quad\quad\quad\quad\quad\quad\quad 1
\end{aligned}
$$

The conclusion can also be reformulated. By P4, 1 is a minimum element in N, and by P2 and the reflexivity clause of P1, 1 is the first element in N. Hence we can restate the conclusion of P5 as:

$$\forall x (x \in N \Rightarrow x \in M)$$

or simply

$$N \subseteq M.$$

Hence, Peirce's mathematical induction axiom, P5, implies that for an arbitrary set M, and the function defined in P6, the following must be the case:

$$1 \in M \wedge M' \subseteq M \Rightarrow N \subseteq M$$

Let A be the family of sets that satisfy the conditions stipulated by the basis and inductive steps:

$$A = \{M \mid 1 \in M \wedge M' \subseteq M\}$$

We know that $N \in A$, since by P4, $1 \in N$, and by our proof of D1, $N' \subseteq N$. We also know by P5 that for every such M, it must be the case that $N \subseteq M$. There is no empty case since N itself is an element of A, so it follows from the definition of intersection that $N = \cap A$.

$$N = \cap \{M \mid 1 \in M \wedge M' \subseteq M\}$$

The right side is just $\cap \{M \mid \{1\} \subseteq M \wedge M' \subseteq M\}$, so by Definition 28:

$$N = \{1\}_{\circ}$$

$\square$

D3. $1 \notin N'$

Proof. This follows directly from P4 and P6. Assume to the contrary that $1 \in N'$. By 25, it must follow that $\exists x(x \in N \land xf1)$, which according to P6 means that $\exists x(x \in N \land 1Sx)$. By definition 19, then, we would have $\exists x(x \in N \land x \neq 1 \land 1Rx)$. But this contradicts P4, and so 1 cannot be a member of N'. □

D4. f is one-to-one.

Proof. This is a consequence of the antisymmetry and connection clauses of P1 and P2. Assume the contrary, that $\exists x \exists y \exists z(xSy \land xSz \land y \neq z)$, and instantiate x, y, and z by the triple a, b, c. It is fairly straightforward to show that, given Peirce's axioms, a contradiction must follow:

i.	$aRb \land a \neq b \land \forall z(zRb \land z \neq b \Rightarrow zRa)$	19 (aSb), H.
ii.	$aRc \land a \neq c \land \forall z(zRc \land z \neq c \Rightarrow zRa)$	19 (aSc), H.
iii.	$a \in N \land b \in N \land c \in N$	i, ii, P0
iv.	$cRb \land c \neq b \Rightarrow cRa$	i, c/z
v.	$bRc \land b \neq c \Rightarrow bRa$	ii, b/z
vi.	$cRb \lor bRc$	iii, P2
vii.	$b \neq c$	H.
viii.	$cRa \lor bRa$	iv–vii
ix.	$\neg cRa$	ii, P1
x.	$\neg bRa$	i, P1

Steps viii - x provide the contradiction, so it must be the case, rather, that

$$\forall x \forall y \forall z(xSy \land xSz \Rightarrow y = z).$$

By P6 and definition 26, it follows that f is one-to-one:

$$\forall x \forall y \forall z(yfx \land zfx \Rightarrow y = z)$$

□

D5. $R = \{(x, y) \mid y \in N \land \{x\} \subseteq \{y\}_{\circ}\}$.

Proof. We must show that, for an arbitrary pair (u, v),

$$uRv \Leftrightarrow v \in N \land \{u\} \subseteq \{v\}_{\circ}.$$

We will first prove the sufficiency, that $uRv \Rightarrow v \in N \wedge \{u\} \subseteq \{v\}_\circ$. That v is an element of N follows immediately from P0. By the proof of D1, we also know that N is a chain, and thus we can use theorems T6–T8 for the unit set $\{v\}$. We will show by mathematical induction starting with v, that every element of N greater than v is also in $\{v\}_\circ$.

(i) $v \in \{v\}_\circ$. Since $\{v\} \subseteq N$ and $N' \subseteq N$, it follows by T7 that $\{v\}_\circ \subseteq N$ and by T8 that $\{v\} \subseteq \{v\}_\circ$.

(ii) $\forall x \forall y(x \in \{v\}_\circ \wedge ySx \Rightarrow y \in \{v\}_\circ$. This is proven by the sequence of steps:

i.	$\{v\}_\circ{}' \subseteq \{v\}_\circ$	T6
ii.	$\forall y(y \in \{v\}_\circ{}' \Rightarrow y \in \{v\}_\circ$	1
iii.	$\forall y(\exists x(x \in \{v\}_\circ \wedge xfy) \Rightarrow y \in \{v\}_\circ)$	25
iv.	$\forall x \forall y(x \in \{v\}_\circ \wedge ySx \Rightarrow y \in \{v\}_\circ)$	iii, P6

Hence, from Peirce's mathematical induction axiom it must be the case that:

$$\forall x(xRv \Rightarrow x \in \{v\}_\circ)$$

But uRv. So $u \in \{v\}_\circ$ and $\{u\} \subseteq \{v\}_\circ$. Hence, $uRv \Rightarrow v \in N \wedge \{u\} \subseteq \{v\}_\circ$.

We next prove the necessity, that $v \in N \wedge \{u\} \subseteq \{v\}_\circ \Rightarrow uRv$. Assume the antecedent, and let M be the set:

$$M = \{x \mid xRv\}$$

To begin, we show that v must be an element of M:

i.	$v \in N$	H.
ii.	$\{u\} \subseteq \{v\}_\circ$	H.
iii.	vRv	i, P1
iv.	$\{v\} \subseteq M$	iii

Now consider an arbitrary element w, in M'. By 25, there must be some element $a \in M$ such that afw. By P6, wSa, and by 19, wRa. But since $a \in M$, we must also have aRv. So by the transitivity clause of P1, wRv. Hence $w \in M$, and because w was chosen arbitrarily from M, it must be the case that $M' \subseteq M$.

v.	$M' \subseteq M$	P1, P6, 25, 19

Since the family of chains containing $\{v\}$ is not empty, we can use T7.

vi.	$\{v\}_\circ \subseteq M$	iv, v, T7
vii.	$\{u\} \subseteq M$	ii, vi, T2

viii. uRv vii

And it follows that $v \in N \wedge \{u\} \subseteq \{v\}_o \Rightarrow uRv$.

Thus, combining both directions, we must have:

$$uRv \Leftrightarrow v \in N \wedge \{u\} \subseteq \{v\}_o$$

$\square$

These proofs show that any structure $< N, R, f, 1 >$ which is a *Peirce natural number system* must also be a *Dedekind natural number system*. Since we have done the derivation in both directions, it follows that a structure $< N, R, f, 1 >$ is a *Peirce natural number system* if and only if it is a *Dedekind natural number system*. Therefore, the two axiom systems are equivalent.

CHAPTER 3

On the Logic of Number

This chapter will provide further historical and philosophical perspective on Peirce's 1881 paper. The first section will examine the mathematical background of this paper; the second section will try to account for its neglect by recent scholarship and to reevaluate its importance for the history of foundations; the third section will deal with its philosophical implications, especially with regard to Peirce's view of the relation between logic and mathematics and how his approach compares with logicism, intuitionism, and formalism.

3.1. Mathematical Background

In order to understand the mathematical background of Peirce's paper it is necessary to take a brief look at the history of the calculus.

The discovery of the calculus introduced into mathematics a host of intellectual problems. It seemed at first almost as if solutions were being obtained by the compounding of errors. Newton's method of integration, for instance, required finding the ratio of increments "infinitely diminished." Newton himself recognized the conceptual difficulty this involved:

> Perhaps it may be objected, that there is no ultimate proportion of evanescent quantities, because the proportion before the quantities have vanished, is not the ultimate, and when they are vanished, is none.

and attempted to answer such objections:

> But by the same argument, it may be alleged, that a body arriving at a certain place, and there stopping, has no ultimate velocity; because the velocity before the body comes to that place, is not its ultimate velocity; when it has arrived, is none.

> But the answer is easy; for by the ultimate velocity is meant
> that with which the body is moved, neither before it arrives at
> its last place and the motion ceases, nor after, but at the very
> instant it arrives; that is that velocity with which the body
> arrives at its last place and with which the motion ceases.[1].

But Newton's response was not convincing. In defining the ultimate ration of evanescent quantities by such phrases as "at the very instant it arrives" and "velocity ... with which the motion ceases," Newton merely reasserted that an infinitesimal can be coherently conceived – which was the very issue in question.

George Berkeley, in his polemic, *The Analyst*, elaborated an objection quite similar to that which Newton had tried unsuccessfully to answer.

> But it would seem that this reasoning is not fair or conclu-
> sive. For when it is said, let the increments vanish, i.e., let the
> increments be nothing, Or let there be no increments, the for-
> mer supposition that the increments were something, or that
> there were increments, is destroyed, and yet a consequence of
> that supposition, i.e., an expression got by virtue thereof, is
> retained ... certainly when we suppose the increments to van-
> ish, we must suppose their proportions, their expressions, and
> everything else derived from the supposition of their existence
> to vanish with them.

He went on to say of Newton that.

> The great author of the method of fluxions felt this difficulty,
> and therefore he gave in to those nice abstractions and geomet-
> rical metaphysics without which he saw nothing could be done
> on the received principles ... It must, indeed, be acknowledged
> that he used fluxions like the scaffold of a building, as things to
> be laid aside or got rid of as soon as finite lines were found pro-
> portional to them. But then these finite exponents are found
> by the help of fluxions ... And what are these fluxions? The
> velocities of evanescent increments. And what are these same
> evanescent increments? They are neither finite quantities, or
> quantities infinitely small, nor yet nothing. May we not call
> them the ghosts of departed quantities?[2]

---

[1]Isaac Newton, *Philosophiae Naturalis Principia Mathematica* [**63**, p. 2]
[2]See [**7**].

Although Berkeley's scruples were clearly influenced by his empiricism, and by his finitistic, conception of geometry,[3] they also represented a concern that the calculus had initiated a decline in the standards of mathematical rigor. "In every other science," Berkeley observed, "men prove their conclusions by their principles and not their principles by their conclusions."[4]

During its first century the calculus prospered, so the question of rigor remained largely in the background. The mathematical fraternity was united, for instance, in its opposition to Berkeley.[5] But by the nineteenth century, a sense of unease was becoming apparent among mathematicians. Technical difficulties appeared, such as the continuous but non-differentiable functions found first by Bolzano in 1830, and later by Riemann in 1854.[6] These functions militated against the naive geometrical conception of the calculus, according to which all continuous functions should be differentiable (at all but possibly a finite set of exceptional points). An even more startling anomaly was Weierstrass' discovery, in 1861, of a function which was everywhere continuous but nowhere differentiable. With the publication of this discovery, by Paul du Bois-Reymond in 1874, the handwriting on the wall had become clearly legible.[7] It would be difficult to sustain a view of the calculus that was based solely upon an intuitive analysis of variation in magnitude.

Because of such difficulties – as well as other researches, e.g., into the representation of functions by Fourier series[8] – mathematicians began to take seriously the sort of criticism expressed by Berkeley. The nineteenth century witnessed efforts to secure the foundations of the calculus, and reestablish rigor in analysis, through a process of arithmetization. This involved, to begin with, generalizations of the notions of "function" and "limit". Through the work of Cauchy and Weierstrass, among others, these notions were gradually divorced from the geometrical intuition that had spawned them.[9] More importantly, the concept of geometrical magnitude itself was subjected to a thorough arithmetization. In 1872 independent treatments of the theory of irrational numbers, by Cantor and Dedekind, made possible the complete reduction of the system

---

[3]See [6] and [8, pp. 84–89].

[4]See [7, p. 30].

[5]See [8, p. xxiii],

[6]See [12, p. 30] and [100]

[7]See [25]. It is interesting to note that nearly half a century elapsed between the first discovery of these pathological functions and their eventual publication – suggesting a hesitance reminiscent of the Pythagorean reluctance to admit the existence of incommensurable magnitudes.

[8]See [63, pp. 34–65].

[9]Ibid, pp. 26–32; also see [17].

of real numbers to that of naturals.[10] The motivation for this theory Dedekind explicitly tied to the need to secure the foundations of differential calculus:

> Even now such resort to geometric intuition in a first presen-
> tation of the differential calculus I regard as exceedingly useful
> ...But that this form of introduction ...can make no claim to
> being scientific, no one will deny. For myself the feeling of dis-
> satisfaction was so overpowering that I made the fixed resolve
> to keep meditating on the question till I should find a purely
> arithmetic and perfectly rigorous foundation for the principles
> of infinitesimal analysis. The statement is so frequently made
> that the differential calculus deals with continuous magnitude,
> and yet an explanation of this continuity is nowhere given,
> even the most rigorous expositions of the differential calculus
> do not base their proofs upon continuity but, with more or less
> consciousness of the fact, they either appeal to geometric no-
> tions or those suggested by geometry, or depend upon theorems
> which are never established in a purely arithmetic manner.[11]

By 1900, the perceived success of the program of arithmetization was such that Poincaré could make the boast that:

> Today there remain in analysis only integers and finite or in-
> finite systems of integers.... Mathematics ...has been arith-
> metized.... We may say today, that absolute rigor has been
> obtained.[12]

Poincaré's claim to absolute rigor might have been premature – the set-theoretical paradoxes would soon enough pose a challenge to the theory of "infinite systems of integers" – but there is a clear sense in which his boast was justified. The techniques of the calculus had been reduced, in effect, to the principles of elementary arithmetic plus set theory.

The arithmetization of analysis was probably the dominant theme of nineteenth century mathematics, and it is important to view Peirce's 1881 paper, "On the Logic of Number," against the background of this process. In a way it was only natural that the attempt to base analysis upon arithmetic should

---

[10]See [17]and [23]. The conceptually distinct, but equivalent, approaches of Cantor and Dedekind are explained in [56, pp. 187–191].

[11]See[pp. 1–2]Dedekind:1872.

[12]Poincaré, Address to the Second International Congress of Mathematicians. See [31, p. 15].

provoke an inquiry into the nature of arithmetic itself. It was no coincidence that three distinct axiom systems for arithmetic were devised within a single decade. The continuity of the development from arithmetization to axiomatization is perhaps most evident in the work of Dedekind, in the transition from his 1872 theory of irrationals to his essay on numbers in 1888.[13]

For Dedekind, though, axiomatization was employed in the service of logicism, i.e., Dedekind subscribed to the logicist attempt to base arithmetic, in turn, upon deductive logic.[14] But not everyone conceived arithmetic to be in the same sort of crisis as analysis. Leopold Kronecker, for instance, embraced arithmetization while denying any need to further secure arithmetic, declaring "God created the natural numbers, the rest is the work of man."[15] And Peirce himself rejected the supposition that arithmetic needed any logical foundation – in this respect being closer to the intuitionists than to the logicists.[16]

Despite this fundamental difference from logicism, Peirce's 1881 program was nonetheless a response to similar historical pressures. Peirce was not unaware of the trend toward arithmetization.[17] He must have realized that the stakes had been raised, that an axiom system for arithmetic had become a matter of greater consequence. His very attention to the question reflected, to some extent, the general sense of crisis which accompanied arithmetization and permeated the mathematical atmosphere of the period. Jerome H. Manheim remarks that:

---

[13]It is instructive to compare the prefaces of the two essays by Dedekind. [**24**, pp. 1–3, 31–40]. That of the latter explicitly refers to the reduction made possible by the former.

[14]*Ibid.* p. 31

[15]See [**10**, p.128].

[16]We will return to this point again in the last section. In one sense, Peirce's rejection of logicism would seem to be a corollary of his position on the nature of inquiry. The logicist reduction, insofar as it was conceived by analogy to arithmetization, was motivated by a sort of methodological doubt. There were no conceptual paradoxes or technical anomalies in arithmetic comparable to those in analysis. Peirce, of course, distrusted such Cartesian distrust: "in cases where no real doubt exists in our minds inquiry will be an idle farce, a mere whitewashing commission which were better let alone." 5.376n3; see also 5.265. On the other hand Peirce clearly recognized the existence of legitimate *philosophical* questions concerning arithmetic.

[17]Peirce was probably unfamiliar with the new treatment of irrationals; he states that he had not read Dedekind's *Stetigkeit* and that he only became acquainted with the work of Cantor in 1884. See 3.563, Ms 27. But his discussion of the doctrine of limits at the end of this decade shows his familiarity with the principles of arithmetization. See 4.118nl and [**27**, pp. 416–421]. Although it is difficult to say just what – or how well – Peirce read, he explicitly cites Lagrange's *Theorie des Fonctions*, an early contribution to arithmetization (*Ibid.,* p. 418), and works by Cauchy and Duhamel on the method of limits (6.125).

> A mathematician working in the eighth decade of the nine-
> teenth century was subject to strains and insecurities rarely, if
> ever, equalled in the previous history of mathematics.... In
> one way or another the mathematical conscience of every ana-
> lyst was threatened.[18]

In this regard, it is important to mention Peirce's adherence, in the face of
arithmetization, to a theory of real infinitesimals. His defense of infinitesimals
should not be interpreted simply as a failure to appreciate the new method of
limits. Murray G. Murphey, for instance, observes that "Peirce's theory was
solidly grounded in the mathematics of his time," and then explains:

> The theory of infinitesimals is rejected today, having been re-
> placed by the theory of limits, but in the 1870's it was still
> widely held by mathematicians. Although Cauchy's works of
> the 1820's had done much to show the superiority of the the-
> ory of limits, it was not until the work of Weierstrass became
> widely known that this superiority was generally acknowledged.
> Meanwhile, many mathematicians of the older generation re-
> mained staunch in their defense of infinitesimals, and among
> the staunchest of these was Benjamin Peirce. To the end of his
> life, Charles never abandoned the theory.[19]

But this is misleading insofar as it suggests that Peirce's view merely elaborated that of an "older generation" which had not kept abreast of mathematical
progress. That Peirce did recognize the mathematical viability of the doctrine
of limits is evident from an 1890 letter written to Simon Newcomb:

> To the doctrine of limits, I have made two criticisms. The first
> is, that the notion that we cannot reason directly about in-
> finitesimals is unfounded. No sane person could conclude from
> that remark that indirect reasoning about them was fallacious.
> The second remark is that the method of infinitesimals "harmo-
> nizes better with recent advances in mathematics." If a reader
> is so thoughtless as to suppose this means the method of limits
> is in downright conflict with recent discoveries in mathematics,
> he will fall into the error you mention. I don't think there is
> much danger of that, but I will change to "is more in the spirit
> of modern mathematical philosophy" in another edition. ...if

---

[18]See [**63**, p. 96].
[19]See [**70**, pp. 119–120].

a limit is correctly defined ... then I see no objection to the
method of limits except its unnecessary circumbendibus'[20]

Peirce conceived the methods of infinitesimals and limits as being, respectively,
direct and indirect ways of doing the same thing. His defense of the former
was not predicated upon rejection of the latter; and appealed not to tradition,
but to "recent advances in mathematics" and to "the spirit of modern mathematical philosophy."[21] In fact, his position – that infinitesimals can provide
an alternative basis for classical analysis – has been cited as anticipating the
development, by Abraham Robinson in 1966, of non-standard analysis.[22]

Peirce himself later abandoned the notion that the method of infinitesimals
"harmonized" better with contemporary mathematical thought, or that it even
dealt with the same subject matter as the method of limits. But he did not
abandon his belief in the mathematical integrity of either approach. If anything,
his mature view left behind the last traces of aspersion ("unnecessary circumbendibus") toward the method of limits. His conception of the real number line
as a *pseudo-continuum*, for example, included the observation that analysis of
functions on real numbers would employ "the universally accepted 'doctrine of
limits'."[23] What Peirce really objected to was not the arithmetical treatment
of such functions, but the assumption that they represented, or were capable of representing, true continuity. In this respect, rather than being "solidly
grounded in the mathematics of his time," Peirce's later defense of infinitesimals
was motivated primarily by philosophical considerations. These considerations
have been discussed elsewhere, as those underlying Peirce's "Kantistic" and
"Post-Cantorian" definitions of continuity.[24] Our point is simply that Peirce's

---

[20]See [**27**, p. 420].

[21]Peirce clearly had in mind Cantor's theory of transfinite numbers, which he believed
would support not only the actually infinite but also the actually infinitesimal. In "The Law
of Mind," an 1892 paper written for *The Monist*, he said that "the word infinitesimal is
simply the Latin form of infinitieth; that is, it is an ordinal formed from infinitum" (6.125),
the implication being that reasoning about such entities will be analogous to that with regard to Cantor's least transfinite ordinal. Although this view extended throughout Peirce's
"Cantorian Period," (See [**91**, pp. 21–25].) he never gave his conception of this analogy much
concrete mathematical content. But Peirce was correct in sensing that the spirit at least
of Cantor's work was more akin to the founders of the calculus than to Berkeley and the
practitioners of arithmetization.

[22]See [**5**, pp. 160–164] and [**104**].

[23]6.176. Peirce also refers to a pseudo-continuum as an "imperfect continuum" (4.642).

[24]See [**91**, p. 27]. The philosophical nature of Peirce's defense of infinitesimals is also
stressed by Benedict in [**5**, pp. 165–177]. Although Murphey appears to make a similar point
later in his book, he suggests a greater degree of competition between Peirce's Kantistic
treatment of continuity and Cantarian continuity than we would prefer – passibly because
Murphey does not distinguish, as consistently as Benedict, Peirce's earlier from his later
views. See [**70**, pp. 281ff]. Also see [**42**].

position on infinitesimals should not be taken as an indication that he was unaware of the mathematical consequences of arithmetization.

For a final link between arithmetization and Peirce's 1881 paper one should look at Carl Boyer's classic, *The History of the Calculus and its Conceptual Development*. Boyer remarks that: "in all probability ... the chief obstacle in the way of the development of the concepts of the calculus was a misunderstanding as to the nature of mathematics."[25] Now Berkeley had already observed, in the eighteenth century, the tendency for analysis to derive principles from conclusions. The later reluctance of analysts to discard the crutch of geometrical intuition was partly a result of this historical emphasis upon applications. Before arithmetization could prevail, the view of mathematics which allowed such an emphasis had to be overcome. Thus the development of the calculus required, in effect, the emergence of a particular conception of mathematics, according to which, as expressed by Boyer:

> Mathematics is neither a description of nature nor an explanation of its operation; it is not concerned with physical motion or with the metaphysical generation of quantities. It is merely the symbolic logic of possible relations, and as such is concerned with neither approximate nor absolute truth, but only with hypothetical truth. That is, mathematics determines what conclusions will follow logically from given premises.[26]

But this notion, that mathematics is purely hypothetical, was Peirce's own view. And, as we shall show in our concluding section, it is precisely this conception of mathematics which "On the Logic of Number" was designed to support. In this sense, Peirce's 1881 paper clearly reflected the spirit of arithmetization.[27]

Aside from the arithmetization of analysis, there was another development in nineteenth century mathematics which provided necessary background for Peirce's axiom system. This was the discovery, by Gauss, Lobatchewsky, and Riemann, of non-euclidean geometry. It is not difficult to see how this discovery, along with researches by Listing, Cayley, Sylvester, and others into topology and projective geometry, must have underlined the general fruitfulness of axiomatics for mathematics.[28] We should at least mention the similarity here between

---

[25][**14**, p. 303].

[26]See [**14**, p. 308].

[27]This paragraph is indebted to [**5**, p. 152–153].

[28]The impact of these geometrical developments upon the resurgence of axiomatics in other areas of mathematics is noted in [**2**, p. 57]. For their impact upon Peirce, see [**70**, pp. 194–228].

Peirce and Hilbert. Both show the influence of these geometrical developments in their overall view of mathematics, in their ready generalization of axiomatic method and thus in their emphasis upon formal systems in mathematics.[29]

3.2. Reevaluation

This section will focus upon the importance of Peirce's 1881 paper for the history of foundations, arguing that it deserves scholarly attention that it has not, in fact, received.

The equivalence of the axiom systems of Peirce and Dedekind can easily be extended to include the well known Peano axioms. To the best of our knowledge, this means that Peirce's 1881 paper was the first in the history of mathematics to contain such an axiom system. Previous philosophers, Leibniz for instance, had pointed out that various arithmetical propositions were logical consequences of more primitive propositions, and many had considered arithmetic to be, in some sense, a deductive science.[30] But Peirce was the first person to ever exhibit a simple set of axioms from which (with a little help from set theory) all arithmetical propositions can be logically derived, and to demonstrate, as well, the actual form such a derivation would take.

It is instructive to compare Peirce's achievement with the most successful prior attempt at axiomatization, by Hermann Grassman in his *Lehrbuch der Arithmetik*, published in 1861.[31] Grassman's *Lehrbuch* provided a group of postulates, not for the natural numbers, but for the integers. Although not originally in axiomatic form, Grassman's characterization of the integers has been given a contemporary formalization by Hao Wang containing the following definitions and axioms:

Definitions
G0. $0 = 1 + -1$
G1. For any a and b, $a - b$ is a number such that $b + (a - b) = 1$
G2. $-a = 0 - a$
G3. $a > b$ iff $(a - b) \in \text{Pos}$
Definitions
G4. $a = (a + 1) + -1$
G5. $a = (a + -1) + 1$

---

[29]See [**70**, p. 187, 235]. See also [**9**].

[30]Leibniz pointed out that, e.g., $3 + 2 = 5$ was a logical consequence of $4 + 1 = 5$, $3 + 1 = 4$, and the general rule that $(a + 1) + 1 = a + 2$. See [**124**, p. 64]

[31]See above, p. 37, note 79.

G6. $a + (b + 1) = (a + b) + 1$

G7. $a \times 0 = 0$

G8. $1 \in \text{Pos}$

G9. $a \in \text{Pos} \Rightarrow a + 1 \in \text{Pos}$

G10. $b = 0$ or $b \in \text{Pos} \Rightarrow a \times (b + 1) = (a \times b) + a$

G11. $b \in \text{Pos} \Rightarrow a \times (-1) = -(a \times b)$

G12. If $1 \in A$, $\forall b, b \in A \Rightarrow b + 1 \in A$, and $b \in A \Rightarrow b + -1 \in A$; then $\forall a, a \in A$.

G13. If $1 \in A$, and $\forall b, b \in A \Rightarrow b + 1 \in A$, then $\forall a, a \in \text{Pos}$ $\Rightarrow a \in A$.[32]

It is important not to underestimate Grassman's system; it anticipated, in many respects, the typical characterization of integers in modern abstract algebra.[33] Nonetheless it is clear that Peirce's axiom system for the natural numbers was conceived on an altogether higher level of abstraction. Furthermore, Grassman's system suffered from two serious defects it did not guarantee the uniqueness of immediate predecessors nor rule out the possibility of positive integers between 0 and 1.

The historical priority of Peirce's axiomatization is in itself sufficient reason for attention to his 1881 paper by scholars in foundations. This priority had been pointed out, during Peirce's own lifetime, by the Dutch mathematician Gerrit Mannoury. In his *Methodologisches und Philosophisches zur Elementar-Mathematik* of 1909, Mannoury stated that.

> A rigorous foundation for the theory of finite numbers, however, was successfully given for the first time by the American C. S. Peirce. This was reproduced in another form or independently rediscovered by others (principally Dedekind, Frege, and Peano).[34]

Mannoury went on to present Peirce's entire axiom system, asserting that his 1881 paper, in general, "is worthy of the highest recognition," and "has been much too little noticed."[35] Mannoury also considered the claim made by Dedekind, in the first preface to *Was sind und was sollen die Zahlen?*, to have

---

[32]See [**123**, p. 148]. Wang's formalization also includes a specification of vocabulary and closure conditions according to which variables, and the results of operations on them, are confined to integers. With general variables G12 would not make sense.

[33]*Ibid.*, p. 147.

[34]See [**64**, p. 51]. The translations throughout this paragraph are our own. We are indebted to Max Fisch for bringing this work to our attention.

[35]*Ibid.*, p. 75

initially conceived his own work in the period 1872–78. In response, Mannoury noted that it was not known when Peirce first discovered his axiom system, and concluded:

> The only question that really deserves the interest of our contemporaries and posterity, however, is that of priority of publication, and this priority without doubt belongs to Peirce.[36]

Since the time of Mannoury, however, Peirce's axiom system has been neglected. Probably the best, and surely the most influential, recent treatment of this topic is Hao Wang's "The Axiomatization of Arithmetic." Wang describes Grassman's attempt, in 1861, to characterize the system of integers. He then gives an excellent account of Dedekind's 1888 essay (from which he believes the Peano axioms to have been "borrowed" – but see above, p. 19, note 40). Wang obviously considers the axiomatization of arithmetic to have been a momentous event in the history of mathematics. He calls Dedekind's axiom system "remarkable," comparing it to Euclid's axiom system for geometry and that of Zermelo for set theory. But he completely overlooks the equivalent axiom system from Peirce's 1881 paper.[37]

Many scholars have cited Wang but have not supplied the chapter that Wang omitted.[38] In fact, the conventional wisdom on this topic is frequently garbled or incomplete – as can be inferred from the following:

> The Italian mathematician Peano was the first to organize the fundamental laws of these numbers in axiomatic form.[39]

> It was not until 1899 [sic] that the arithmetic of cardinal numbers was axiomatized, by the Italian mathematician Giuseppe Peano.[40]

---

[36]*Ibid.*, p. 78

[37][**123**, pp. 145, 153].

[38]Wang is cited, for example, in [**31**, pp. 102, 292–293], [**97**, p. 92], and various other works. Quine's reliance on Wang is especially surprising in view of the fact that Quine had written a review of the *Collected Papers*, in 1934, which acknowledged Peirce's 1881 "systematization of the arithmetic of positive integers." [**95**, p. 294].

[39]See [**2**, p. 58].

[40]See [**71**, p. 103].

The first semi-axiomatic presentation of this subject was given
by Dedekind (1901) and has come to be known as Peano's
postulates.[41]

This formulation is due essentially to Peano, who first formulated a postulational basis for the counting numbers ... [42]

The origins of both the abstract method and the critical approach can be traced definitely to the 1880's ... it seems just to
attribute the initial impulse to G. Peano ... in his postulates
for arithmetic (1899).[43]

This sampling is representative, and extension of it would merely reinforce the
impression that the axiomatization of arithmetic originated with Dedekind and
Peano. As near as we have been able to determine, the priority of Peirce's 1881
axiom system has not been recognized by any recent work on the foundations or
history of mathematics.[44] Peirce scholarship has recognized his emphasis upon
axiomatization, but not the significance of his 1881 axiom system.[45]

Peirce is generally conceded a crucial role in the development of modern
logic. Even among his contemporaries, Peirce's reputation as a logician was

---

[41]See [**65**, p. 102]. The date in parenthesis, here, is actually a citation of Beman's
translation of the Dedekind essay. Mendelson also cites Wang's "Axiomatization."

[42]See [**54**, p. 142].

[43]See [**3**, p. 244].

[44]There are a few cryptic references to Peirce's axiom system in Beth's *Foundations*,
[**10**, pp. 113, 360]. Together with that in Quine's 1934 review [**95**, p. 294], these are
the only references of any sort to Pierce's axiom system since the time of Mannoury – at
least that we have managed to discover. Most of the other works cited here do not even
mention Peirce's 1881 axiom system, much less its equivalence to the Dedekind–Peano axioms.
Two recent articles in *The Encyclopedia of Philosophy*, "Logic, History of" by A. N. Prior,
and "Mathematics, Foundations of" by Charles Parsons, refer to the contributions of both
Dedekind and Peano, but not to that of Peirce.

[45]Murphey, for instance, notes that Peirce constructed an axiom system from which
the Peano postulates can be derived. But this has not helped the case for Peirce's priority
since the system to which he refers (3.562ff. – intended for inclusion in Peirce's "The Logic
of Mathematics in Relation to Education" [**81**]) did not appear until a full decade after the
Dedekind-Peano axioms, and nearly two decades after Peirce's own 1881 system [**70**, pp. 243–
244]. Similarly, Goudge cites an axiom system which was sketched out by Peirce around 1897
(4.160) without apparently realizing that Peirce had already published the axiomatization in
"On the Logic of Number." See [**38**, p. 67]. We are not aware of any contribution to Peirce
scholarship which has yet realized the significance of Peirce's 1881 paper.

widespread.[46] Recent scholarship has reinforced this reputation, attributing to Peirce such major technical contributions as:

1. improving Boole's + operation by making it inclusive
2. pioneering the study of the logic of relations
3. introducing into logic the use of quantifiers and bound variables
4. using the calculation of truth-values to establish logical laws
5. axiomatizing the propositional calculus
6. initiating the study of 3-valued logics
7. anticipating aspects of the logic of strict implication and the semantics of modal logic
8. discovering the expressive completeness of the Sheffer connective and the Peirce connective
9. originating the method of natural deduction.[47]

A. N. Prior states that "there is probably no logical writer who has been more rich in original suggestions than Peirce, and his papers are a mine that has still to be fully worked."[48]

In contrast, the more mathematical side of Peirce's work has been far less visible. Nothing has been written, for instance, from which it would be possible to cull a comparable list of Peirce's contributions to set theory and axiomatics.

---

[46]E.g., Ernst Schröder, whose epochal *Vorlesungen Über die Algebra der Logik* [113] systematized the Boolean algebra and paved the way for the modern period, was greatly influenced by Peirce and corresponded extensively with him; Giuseppe Peano cited Peirce's logical work in the preface to *Arithmetices* [53, p. 102]; Bertrand Russell often referred to Peirce's work on logic in his early writing [109, pp. 23, 26 *passim*]; Peirce's logical reputation was further enhanced by the work of two of his students at Johns Hopkins, O. H. Mitchell and Christine Ladd-Franklin, who made important contributions to Peirce's *Studies in Logic by Members* of the Johns Hopkins University [79]; as a young man Peirce had exchanged letters with Augustus De Morgan, and we have later drafts of letters to Jevons and MacColl, including a reply from the latter; Carolyn Eisele cites a letter written by one E. L. Youmans to his sister, dated London, October 29, 1877: "Charles Peirce isn't read much on this side. Clifford, however, says he is the greatest living logician, and the second man since Aristotle who has added to the subject something material, the other being George Boole, author of The Laws of Thought." [86, vol. 1, p. xxii]

[47]For a general treatment of Peirce's contributions to logic, see [59] and [92]. (1) 3.1-3.19; this was done independently of Jevons who had made a similar improvement three years earlier. (2) Although this is often credited to De Morgan, Emily Michael [66] makes a convincing case that Peirce independently discovered the logic of relations. Peirce, at any rate, developed the field almost singlehandedly. (3) 3.328-3.357, independently of Frege, but with some debt to his student O. H. Mitchell. See 3.363, 3.393. (4) Prior to Frege, see [119, p. 95] (5) 3.376-3.384, independently of Frege. See *ibid.,* p. 101. (6) See [30]. (7) See [92] and [93]. (8) 4.12-4.02, written 33 years before [114]. (9) See [101].

[48]See [93].

This is partly due to the circumstances of Peirce's life. The main period of his interest in set theory occurred later in his life, toward the end of his career. Yet after leaving Johns Hopkins in 1881, Peirce did not hold another academic appointment. His mathematical work in the ensuing decades was produced mainly in isolation, in a house on the Delaware river, lacking the institutional affiliation and intellectual community which might have supported and helped promulgate his ideas. As a result, Peirce's mathematical writing has been less accessible than his earlier logical work.

But this does not explain why the neglect of Peirce's mathematical work should extend into the latter part of the twentieth century. To understand this, we must look to a conventional picture of nineteenth century logic. This picture involves the distinction between such figures as Boole, De Morgan, Jevons and Schröder, who were interested primarily in the development of logic, and Frege, Dedekind, Cantor and Peano, who employed the new logic as a tool for research into the foundations of mathematics. Peirce is often associated with the former group, since it comprised the historical context for his early work on logic. But to the extent that this division has become a kind of entrenched wisdom, it has helped to obscure Peirce's more mathematical work – including his 1881 axiom system.[49] Peirce does not fit well into this conventional picture.

Another problem with the conventional picture is that it can encourage an unwarranted judgment with regard to the historical importance of these figures. Jean van Heijenoort, for example, asserts that the Boolean period was not a "great epoch" in the history of modern mathematical logic, and says that such an epoch began only in 1879, with the publication of Frege's *Begriffsschrift*. [50] But this assessment is surely mistaken: Gregory Moore notes that it radically underestimates the contributions of Boole and Schröder.[51] Moreover, Peirce's pivotal role in the development of mathematical logic is completely ignored – it has apparently fallen into an interstice between epochs.[52]

By establishing the equivalence of Peirce's 1881 axiom system to the well known Dedekind–Peano axioms we have provided a straightforward argument for reevaluating Peirce's contribution to the foundations of mathematics. We have also mentioned various technical contributions in Peirce's 1881 paper, some

---

[49]See [92], [59, pp. 12–18], [117, p. 1], and [10, p. 64]. In general Beth is an exception to this tendency - (*ibid.*, pp. 67, 113, 360) because of the concreteness of his approach and, perhaps, the direct influence of Mannoury.

[50]See [44, p. *vi*].

[51]See [69, p. 469].

[52]Carolyn Eisele and Murray Murphey have sought to correct this picture by stressing the importance of Peirce's more mathematical work. See [70]. A further list of Eisele's articles is given at the end of [29].

of which have been previously noted in the literature, some not. Among the former are:

1. using recursive definitions for arithmetic operations
2. the first known definition of a Dedekind-infinite set.

While the latter include:

3. a corresponding definition of an ordinary infinite set (without assuming the axiom of choice)
4. the first ordinal construction of cardinals
5. the first abstract description of partially and simply ordered sets.[53]

Peirce's axiom system is not without theoretical interest. Although unlikely to replace the practical Peano axioms, it is surprisingly economical: a simply ordered set, closed with respect to predecessors, having a minimum but no maximum, and obeying mathematical induction. An important technical feature of this system is its statement of the axiom P3. If Peirce had formulated this axiom so as to assert closure with respect to successors, then it would have guaranteed, in conjunction with P1 and P2, that his numbers were well ordered. But this was unnecessary, since well-ordering is provided in any case by the addition of P4 and P5. And by requiring closure, instead, with respect to predecessors, Peirce anticipated the technique recently advocated in Quine's *Set Theory and its Logic*. Quine describes this approach as follows:

> A familiar definition of natural numbers involves infinite classes. Natural numbers are the common members of all classes that contain 0 (somehow defined) and are closed with respect to the successor operation (somehow defined) and any such class is infinite. The law of mathematical induction, based on this definition, can be proved only by assuming infinite classes. But I get by with finite classes by inverting the definition of natural number, thus. x is a natural number if 0 is a common member of all classes that contain x and are closed with respect to predecessor.[54]

Peirce was not afraid of infinite classes. But his formulation of P3 allowed his axiom system to focus its supposition of such classes on the axiom P4. Thus a simple modification – deleting the requirement that N not have a maximum –

----

[53]See Chapter 1, pp. 6–7, 22–27.
[54]See [**97**, pp. 75–77, 83ff].

would remove this supposition without destroying the sense of Peirce's axiom system. It would leave a system which, like that of Quine, would serve just as well if the natural numbers happened to be finite. No such simple modification is possible with the systems of Dedekind or Peano.

In short, despite being Peirce's first real contribution to this field, "On the Logic of Number" provides, by itself, a compelling case for reconsidering Peirce's role in the early history of foundations.

3.3. Peirce on the Relation between Mathematics and Logic

The idea of axiomatization leads naturally into questions about the relation between logic and mathematics. This section will explore Peirce's mature understanding of this relation. We will look at the immediate philosophical context of Peirce's 1881 paper and explicate his conception of mathematical reasoning by contrasting it to the traditional triad of logicism, intuitionism, and formalism.

3.3.1. Philosophical Context.
There are two points which need to be made with regard to the immediate philosophical context of Peirce's 1881 paper. The first is referred to, although somewhat obliquely, in Peirce's own introduction. Peirce began his essay by saying.

> Nobody can doubt the elementary propositions concerning
> number: those that are not at first sight manifestly true are
> rendered so by the usual demonstrations. But although we
> see they *are* true, we do not so easily see precisely *why* they
> are true; so that a renowned English logician has entertained
> a doubt as to whether they were true in all parts of the uni-
> verse. The object of this paper is to show that they are strictly
> syllogistic consequences from a few primary propositions. The
> question of the logical origin of these latter, which I here regard
> as definitions, would require a separate discussion. (3.252)

The "renowned English logician" referred to in this passage is surely John Stuart Mill. Mill had proposed a radical empiricism which, in contrast to the classical empiricism of Locke, Berkeley, and Hume, included even mathematics. Mill held that particular arithmetical propositions, such as $2+2=4$, were inductive generalizations from observed physical facts. Although highly confirmed, such propositions, according to Mill, did not differ in kind from the generalizations of empirical science. Thus arithmetic became, for Mill, a sort of *synthetic a posteriori* knowledge, derived directly from experience. And if our experience were different, e.g., in another part of the universe, then our arithmetic might require adjustment as well.

Mill's view is not very common today. Evert Beth, for instance, remarks that "the empiricist view of mathematics, as defended by John Stuart Mill, is of so little influence in the present day that it may be left out of consideration ..."[55] But it is important not to project this lack of influence back onto the philosophical milieu of Peirce's 1881 paper. During the middle decades of the nineteenth century Mill's view was much more widely held. His *A System of Logic*, for instance, was used as a standard text at both Oxford and Cambridge.[56] Peirce, who first studied Mill's *Logic* at Harvard and later taught a course on it at Johns Hopkins, commented upon the popularity of Mill's doctrines, "which many persons adhere to passionately without reference to their meaning." (4.33) And Frege evidently considered Mill's empiricism a force to be reckoned with, judging from the space he devoted to refuting it in his *Grundlagen* of 1884.[57]

In his 1881 paper Peirce clearly set out to contest Mill's empiricism by proving that arithmetical truths "are strictly syllogistic consequences from a few primary propositions." His argument does not require much elaboration: if arithmetic can be axiomatized, then individual theorems will be deductive consequences of these axioms rather than highly confirmed inductions from experience. The very success of Peirce's program, and of similar efforts by Frege, Dedekind, and Peano, has tended to obscure the philosophical antecedents of these achievements. Thus we find it difficult to imagine, today, an intellectual environment in which the empiricism of Mill was a defended, if not defensible, position in the philosophy of mathematics. But such an effort is necessary if one is to appreciate the historical significance of Peirce's axiom system.

The second point involves the facts of publication of Peirce's 1881 paper. "On the Logic of Number" was published the year after the death of Peirce's father, Benjamin Peirce, in 1880. It immediately preceded, in the same issue of *The American Journal of Mathematics* the posthumous publication of the elder Peirce's *Linear Associative Algebra,* which began with his famous definition of mathematics as "the science which draws necessary conclusions."[58]

---

[55]See [**11**, p. 3].

[56]See [**67**]. Mill's views on mathematics were given in the first two chapters of this work. On its wide dissemination see Encyclopedia of Philosophy, s.v. "Mill, John Stuart," by J. B. Schneewind.

[57]See [**33**, pp. 9–17, 29–33].

[58]See [**76**, p. 97].

Whatever incidental reasons there might have been for this juxtaposition, Peirce's 1881 program was clearly related conceptually to his father's definition. If mathematics indeed draws necessary conclusions, then there ought to be an axiom system for arithmetic, i.e., a set of propositions *from which* the conclusions of arithmetic are drawn. Conversely, in view of the arithmetization of analysis, such an axiom system ought to provide the "missing link" in support of Benjamin Peirce's definition, making it possible to exhibit the entire range of reasoning underlying the conclusions of analysis.

It is hardly conceivable that Peirce was not aware of this natural affinity between his 1881 paper, with its attempt to show that arithmetic consisted of "strictly syllogistic consequences from a few primary propositions," and his father's celebrated definition of mathematics. Their direct juxtaposition seems to imply such an awareness. If so, then it provides an important clue to how Peirce himself might have formulated the broader philosophical implications of his 1881 paper – since Peirce eventually incorporated his father's definition into his own philosophy of mathematics and, in particular, used it to express his conception of the relation between mathematics and logic.

3.3.2. Logicism.
Logicism conceived mathematics to be an extension of deductive logic. Thus the logicist program, as formulated by Frege, Dedekind, and the early Russell, was to exhibit the necessary definitions and the precise logical steps by which the theorems of classical mathematics could be derived from the laws of logic.[59]

There is no question but that in his later writing Peirce rejected logicism. (2.81. 5.126) He interpreted his father's definition to imply that mathematics, the science *which draws* necessary conclusions, is essentially different from deductive logic, the science "*of drawing* necessary conclusions." (4.239) And he used this interpretation to distinguish his own conception of mathematics from the logicism of Dedekind which, he said, "would not result from my father's definition."[60]

---

[59]See [**56**, pp.33–51].

[60]4.239. emphasis added. See [**24**, p. 31]. Historically there have been various other interpretations of Benjamin Peirce's definition. Hermann Weyl, for example, understood it to mean that and distinction between logic and mathematics had been "obliterated." See [**124**, p. 62]. But there is no evidence that Peirce ever interpreted this definition in other than the manner indicated above. He even claimed that this was the interpretation his father had originally intended:

> "It is evident, and I know as a fact, that he had this distinction [between logic and mathematics] in view. At the time when he thought out this definition, he, a mathematician, and I, a logician, held daily discussions about a large subject which interested us both; and he was struck, as

On the other hand, it is not immediately clear when Peirce first adopted this definition, or when he first arrived at his mature position on the distinction between mathematics and logic. Thus there is some suggestion, in the literature, that Peirce might have subscribed to logicism in his earlier writing.[61] We will look a bit further into the evidence on this matter.

Peirce's 1881 paper, despite its opposition to Mill's empiricism and its typically logicist technical objective, does not seem to have been conceived from a logicist philosophical perspective. This is implied by its obvious juxtaposition to his father's definition of mathematics. It is substantiated by Peirce's explicit refusal, in his introduction, to commit himself to the derivation of arithmetic from logic. The question of the logical origin of his axioms, he stated, "would require a separate discussion." (3.252) Furthermore, as was pointed out in our first chapter, the full logicist program included the attempt to assure a model for the natural numbers. Yet it was concerning precisely this attempt that Peirce's 1881 paper differed most visibly from *Was sind und was sollen die Zahlen?*.[62] On balance, there is little reason to suppose that in 1881 Peirce sympathized with logicism.

In terms of its overall program, Peirce's 1881 paper seems to have been patterned after his "Upon the Logic of Mathematics," published in 1867.[63] This earlier paper had also aimed to axiomatize arithmetic, and contained an introduction remarkably similar to that of his 1881 paper:

> The object of the present paper is to show that there are certain general propositions from which the truths of mathematics follow syllogistically. and that these propositions may be taken as definitions of the objects under the consideration of the mathematician without involving any assumption in reference to experience or intuition. That there actually are such objects in experience or pure intuition is not itself a part of pure mathematics. (3.20)

The technical objective, of showing that "the truths of mathematics follow syllogistically" from "certain general propositions," is almost identical to that proposed in "On the Logic of Number." The main difference would appear to

---

I was, with the contrary nature of his interest and mine in the same propositions." (4.239)

[61]See below, pp. 118–121.
[62]See Chapter 1, pp. 56–70.
[63]See [**77**], reprinted at 3.20–3.44.

be that in 1867 Peirce was careful to divorce this task from "any assumption in reference to experience or intuition," while in 1881 his corresponding reservation concerned "the question of the logical origin" of his axioms.

Thomas Goudge interprets Peirce's earlier 1867 paper as being in the spirit of logicism. He says:

> In the passage already quoted from the opening part of his paper "Upon the Logic of Mathematics," it is clear that he thought of the science as resting on a foundation of logical definitions. Indeed, in the last section of the paper there is a definite foreshadowing of the theme later elaborated by Frege in his *Grundlagen der Arithmetik* (1884) and by Whitehead and Russell in *Principia Mathematica* (1910–13), that arithmetic is derivable from logic.[64]

Goudge goes on to observe that Peirce's 1867 paper "belongs to the general tradition of Cantor and Frege."[65] Since Goudge is aware that Peirce's mature view was not logicist, he claims that Peirce "repudiated his youthful attempt to derive mathematical notions from logic."[66] Accordingly, Goudge tries to account for this apparent "shift in doctrine," and to explain what caused Peirce to "reverse his position." He concludes that the principal factor in this change must have been Peirce's acceptance of his father's definition of mathematics.[67]

Neither Goudge's interpretation of Peirce's 1867 program, nor his claim that Peirce subsequently reversed his position, is very convincing. There are a number of difficulties: First, the similarity of this program to that adopted in 1881 provides a strong prima facie argument for the continuity of Peirce's underlying conception of mathematics. Goudge appears to conflate Peirce's intention to axiomatize arithmetic, which was indeed shared by logicism, with the full logicist program of deriving mathematics from logic. The latter required not only the axiomatization of arithmetic, but also the derivation of the axioms themselves from logic. In both of his papers, however, Peirce proposed *only* axiomatization – and there is no reason to consider axiomatization, in itself, an exclusive characteristic of logicism. Second, Peirce's 1867 program was also similar to "On the Logic of Number" in avoiding the question of the existence of a model. "That there actually are such objects," Peirce said, "... is not itself a part of pure mathematics." Third, Goudge suggests that the last section of

---

[64]See [**38**, p. 57].

[65]*Ibid.*, p. 66

[66]*Ibid.*, p. 65

[67]*Ibid.*, p. 57

Peirce's 1867 paper gives additional evidence that Peirce was a logicist. But this final section merely gives Peirce's anticipation, cited in our first chapter, of the Frege-Russell abstractive definition of cardinals.[68] We can see no reason why it should be construed as implying a commitment to logicism, any more than should Peirce's 1881 ordinal definition of cardinals. It certainly does not involve any "foreshadowing of the theme ... that arithmetic is derivable from logic," especially since, as Russell pointed out, arithmetic has very little to do with the cardinal properties of numbers anyway.[69] Fourth, even if Peirce were a logicist in 1867, Goudge has not satisfactorily accounted for what would have led him to change his position. Judging from Peirce's own recollection, he already had serious reservations about logicism before his father's definition was even formulated,

> The only way in which I think that anything I said influenced anything in my father's book (except that it was partly on my urgent prayer that he undertook the research) was that when at one time he seemed inclined to the opinion which Dedekind long afterward embraced, I argued strenuously against it, and thus he came to take the middle ground of his definition.[70]

Goudge's interpretation requires Peirce to have changed his thinking sometime between 1867 and 1870 (when Benjamin Peirce's *Linear Associative Algebra* was first circulated privately. See 4.301nl), but fails to provide an adequate motivation for such an abrupt change. Considering as well the other difficulties with Goudge's account, the simpler hypothesis would appear to be that Peirce did not change his position at all.

Finally, in support of his interpretation Goudge also cites the harsh judgement Peirce later made of his 1867 paper: "It is now utterly unintelligible to me, and is. I trust, by far the worst I ever published." (4.333) Goudge remarks that, "Even making allowance for the rhetorical flourish of this statement, it strongly suggests a shift in doctrine."[71] But Peirce's later judgement surely referred to the implementation of his 1867 program rather than to this program itself. As we pointed out in Chapter 1, Peirce's 1867 paper utilized a primitive Boolean calculus which Peirce then discarded upon developing the logic of relatives. Thus his disavowal indicated not a shift in doctrine, as Goudge would have it, but an improvement in technique. Moreover, Peirce never expressed dissatisfaction with the almost identical program of his 1881 paper. And his

---

[68]See Chapter 1, p. 46.

[69]See Chapter 1, p. 41.

[70]"Notes on B. Peirce's *Linear Associative Algebra*," Ms 78 cited in [**70**].

[71]See [**38**, p. 55].

later writing shows that he continued to attach great significance to the axiom-atization of arithmetic. In volumes III and IV alone of the *Collected Papers* there are at least ten separate presentations of such an axiom system.[72]

In short, there appears to be no justification for Goudge's notion that Peirce was a repentant logicist. As noted above, there was only one appreciable differ-ence between Peirce' s 1867 and 1881 programs – and this difference concerned how Peirce feared he might be misinterpreted, not how he wished to be inter-preted. Thus his 1867 caution not to construe his axioms as involving "any assumption in reference to experience or intuition" should be understood in terms of the general Kantian context of his early thought.[73] In 1881 Peirce was more aware of the possibility that his technical program might be confused with logicism, hence the corresponding disclaimer about "the question of the logical origin" of his axioms. From 1867 on, at any rate, there is no reason to suppose that Peirce ever seriously entertained logicism. He did not consider mathemat-ics to be an extension of, or reducible to, deductive logic. But the question remains as to how Peirce did understand the relation between mathematics and logic. Of the traditional approaches to the philosophy of mathematics, his position, on this particular issue, was probably closest to that of intuitionism.

3.3.3. Intuitionism.
Intuitionism, in its modern form, was the creation of the Dutch mathematician L. E. J. Brouwer.[74] Although he followed many nineteenth century mathematicians in rejecting the Kantian notion of the a pri-ority of (Euclidean) space, Brouwer firmly maintained the a priority of time, and held that time provided a basal intuition of "one-twoness" out of which arith-metic and analysis could be developed. Since he conceived this development to involve "essentially languageless" mental constructions, Brouwer also held mathematics to be independent of, and prior to, any mathematical language – including that of theoretical logic:

> The first act of intuitionism completely separates mathematics
> from mathematical language, in particular from the phenom-
> ena of language which are described by theoretical logic, and
> recognizes that intuitionist mathematics is an essentially lan-
> guageless activity of the mind having its origin in the percep-
> tion of a move of time, i.e., of the falling apart of a life moment
> into two distinct things, one of which gives way to the other,
> but is retained by memory.[75]

---

[72]Viz., 3.253ff., 3.562, 4.110, 4.160ff., 4.188, 4.335, 4.336, 4.341, 4.606, 4.664ff.
[73]This could also include a reference to Mill.
[74]See [**56**, pp.118–134].
[75]See [**15**, pp. 140–141].

Brouwer's view was further elaborated by the intuitionist Arend Heyting, who claimed that logic is actually dependent upon mathematics – being a generalized description of mathematical practice:

> If the construction P shows that $A \Rightarrow B$, and Q shows that $B \Rightarrow C$, then the juxtaposition of P and Q shows that $A \Rightarrow C$. We have obtained a logical theorem. The process by which it is deduced shows us that it does not differ essentially from mathematical theorems; it is only more general, e.g., in the same sense that "addition of integers is commutative" is a more general statement than "$2 + 3 = 3 + 2$." This is the case for every logical theorem, it is but a mathematical theorem of extreme generality; that is to say, logic is a part of mathematics, and can by no means serve as a foundation for it.[76]

In this sense, intuitionism completely reversed the logicist conception of the relation between mathematics and logic, This reversal, together with. the fact that intuitionist mathematics was itself somewhat unusual, accounted for what is probably the best known feature of intuitionism, viz., its rejection of the law of excluded middle.[77]

Before we continue, it is important to distinguish Peirce's treatment of the specific issue of how mathematics is related to logic from his overall conception of mathematics. There are, indeed, several other features of Peirce's thought which might appear to resemble certain tenets of intuitionism. He questioned the universal applicability of the law of excluded middle (1.434, 4.640, 6.182), and the standard interpretation of quantification (6.121, 6.165). His pragmatism might be misunderstood as having intuitionist overtones.[78] And it has been suggested that Peirce's thought shows an intuitionist emphasis upon the observational and experimental aspects of mathematics.[79]

---

[76]See [**48**, p. 6].

[77]Intuitionist mathematics differed from classical mathematics in being both more rigorous and more restrictive, It conceded mathematical existence only to actual mental constructions, Thus intuitionism understood $\exists x$ to mean: "I have effected the mental construction of an x such that _____," It also interpreted negation constructively, so that $\neg\exists x$ became: "from the supposition that there is an x such that_____, I have constructed a contradiction." It is easy to see that, for an intuitionist, $\exists x(Fx) \vee \neg\exists x(Fx)$ must fail whenever the determination of an F remained undecided, For example, we can presently assert neither that we have constructed the greatest prime twin, nor that from the supposition of such a twin we have constructed a contradiction. *Ibid.*, pp. 2–3, 101ff.

[78]See [**74**, pp. 243–253].

[79]See [**38**, pp. 55–57]. Arthur Burks is also cited in this regard by Murphey [**70**, p. 230].

Nonetheless, the main thrust of Peirce's mathematical thought was clearly *not* intuitionist. Peirce was not a constructivist. In contrast to intuitionism, which confined itself to mental constructions and thus sought to expurgate from mathematics all reference to the actual infinite, Peirce accepted the entire Cantorian hierarchy of transfinite numbers, and considered the chief value of Cantor's work to consist precisely in demonstrating that the infinite is amenable to mathematical reasoning. (See 6.113–4, 8.117) Although Peirce might have entertained intuitionism as a sort of "mathematics of secondness," or as an exclusively hypothetical undertaking, he would not have sympathized with the broader philosophical pretensions of intuitionism. To the extent that it adopted constructivism as a dogma, Peirce would have regarded intuitionism an impediment to mathematical inquiry.

Peirce's own reservations about the law of excluded middle stemmed not from any constructivist scruples but from the difficulty in extending this law to generals and possibles:

> Now if we are to accept the common sense idea of continuity (after correcting its vagueness and fixing it to mean something) we must either say that a continuous line contains no points or we must say that the principle of excluded middle does not hold of these points. The principle of excluded middle only applies to an individual (for it is no.t true that "Any man is wise" nor that "Any man is not wise"). But places, being mere possibles without actual existence, are not individuals.[80]

Peirce did not reject the law on prior epistemic grounds as did the intuitionists.[81]

Peirce also warned specifically against a constructivist misinterpretation of his pragmatism;

---

[80]6.168. See also 5.448, 5.505. Peirce explicitly tied his doubts about the law of excluded middle to the problem of modality, "a matter that I cannot pretend to have got to the bottom of; and logic here seems to touch metaphysics" (6,182), in this sense seeming to anticipate the correlation discovered by Gödel in 1932, between intuitionist logic and the modal system S4.

[81]Thus Peirce also spoke of infinite sets to which the law of excluded middle does apply. holding only that individuality decreases with size (Kantistic period. see 4.172–4.175. 4.188ff., 6.185–6.186). Like the syllogism of transposed quantity and mathematical induction, the law of excluded middle becomes a casualty of the size of collection under consideration (Kantistic period), or the difference in dimensionality between such a collection and its elements (Post-Cantorian period).

> ...the principle [pragmatism] might easily be misapplied to sweep away the whole doctrine of incommensurables, and, in fact, the whole Weierstrassian way of regarding the calculus. (5.3)

What bothered Peirce was the naive operationalist interpretation of pragmatism, according to which, e.g., sufficiently large decimal places become meaningless because they have no immediate bearing upon action or perception. Peirce rejected any such interpretation, explaining that his pragmatic maxim had intentionally been formulated to "avoid all danger of being understood as attempting to explain a concept by percepts, images, schemata, or by anything but concepts." (5.402n3) And Peirce certainly did not believe that thought could be reduced to action, any more than that thirdness in general could be reduced to secondness. He would have measured the significance of a concept like incommensurability in terms not only of its effect upon immediate behavior, but also of its possible bearing upon our "conceived conditional resolutions" to act within a highly contingent future.[82]

Thus a correct application of Peirce's pragmatism would permit both classical and constructivist mathematics, and distinguish between them on the basis of consequences so general and conceptual as to be summed up by Peirce in the phrase "rational purport."[83]

---

[82]See 5.453. That Peirce considered mathematics itself a first, and purely hypothetical, is a separate issue which will be further discussed below. Peirce clearly thought his pragmatism allowed scope for the infinite:

> the connection of the doctrine (of multitude) with the logical maxim called pragmatism is interesting. All things that exist ought by pragmatism (as a regulative principle) to form an enumerable collection. But what may be in futuro forms a denumerable collection. Now, according to tychism, law determines some things (excludes some future contingencies) and leaves others indeterminate. In that case the possible different courses of the future have a first abnumerable multitude.

Draft of a letter to E. H. Moore. Dec. 29. 1903. Ms L299. cited in [29, pp. 232–233].

[83]5.428. Our interpretation of Peirce's pragmatism follows Potter [90, pp. 52–67]. Patin is surely mistaken in interpreting Peirce's formulation of the pragmatic maxim as necessarily constructivist. After citing Peirce's description of a logical interpretant,

> "...to predicate any such [intellectual] concept of a real or imaginary object is equivalent to declaring that a certain operation, corresponding to the concept, if performed upon the object, would ...be followed by a result of a definite general description."(5.483)

Patin says that "this should have been the birth of 'operationalism'," and that "Peirce's statement ...is also essentially the position of ...'intuitionists'" [74, p. 244]. Yet Patin ignores the disclaimer with which Peirce began the very next paragraph, viz., that this description is still incomplete, and Patin apparently misses the important distinction Peirce emphasized

In this respect Peirce's pragmatism was different from that of the later Wittgenstein.[84] Wittgenstein tied his own criterion – "meaning is use" – more closely to actual experience and immediate behavior. Accordingly, he also accepted the implications of this criterion for his philosophy of mathema:tics, which was constructivist and strictly finitistic.[85] Peirce's pragmatic maxim, on the other hand, was formulated in terms broad enough to encompass his realism – and Cantor's paradise.

Finally, it is true that Peirce often stressed the importance of observation and experiment in mathematics. The following passage is typical:

> But how does this evolution of necessary consequences take place? We can answer for ourselves after having worked a while in the logic of relatives. It is not by a simple mental stare, or strain of mental vision. It is by manipulating on paper, or in the fancy, formulae or other diagrams – experimenting on them, *experiencing* the thing. Such experience alone *evolves* the reason hidden within us ... (4.86)

On the basis of this passage Goudge attributes to Peirce a sort of intuitionist epistemology.[86] But while there are indeed aspects of Peirce's thought that resemble intuitionism (see below), his emphasis upon observation and experiment are not necessarily among them. Goudge is certainly mistaken, for instance in holding that this passage requires a "'formal intuitive induction'" to account for the movement from particular formulae and diagrams to general mathematical assertions.[87] Peirce did not understand the inference from mathematical diagrams as analogous to scientific induction from particulars to generals. He conceived it more on the model of the logical inference from arbitrary parameters to quantified variables. The mathematical diagram is an *icon*, a sign which

---

later in this same paper, between an "energetic interpretant" and a "logical interpretant." (5.486, 5.491, 5.494)

[84]If there was any historical influence of Peirce's pragmatism upon Wittgenstein, it was probably through the agency of F. P. Ramsey. One of the rare direct citations in Wittgenstein is of Ramsey quoting Peirce.[**126**, p. 38E] Ramsey's own admiration for Peirce is evident from his Essays [**98**, pp. 96ff., 149].

[85]The constructivist implications of Wittgenstein's identification of meaning and use are examined by John Richardson [**99**, pp. 45–77]. Richardson claims that Wittgenstein's return to philosophy was in fact precipitated by a lecture given by Brouwer in Vienna, in March, 1928 (*Ibid.*, pp. 11–17). Also see [**26**].

[86]Goudge says that this passage portrays mathematics as "an observational and experimental science which deals with diagrams, or signs, created by the human mind," and that it implies a special intuition by which mathematical demonstrations are justified.[**38**, pp. 56, 57].

[87]*Ibid.*, p. 57

represents a form, a sign which has already been prescinded from extraneous empirical content (4.531). This is why Peirce was able to say that manipulation of diagrams can evolve *necessary* conclusions. In fact, Peirce did not draw a line between mathematics and deductive logic with regard to their use of diagrams – referring in this passage only to his work "in the logic of relatives."[88]

Goudge's confusion can be traced to the fact that he overlooks the important distinction with which Peirce introduced this passage, viz., between questions of psychology and questions of logic.[89] Peirce was attacking a *psychological* conception of analyticity – Kant's notion that an analytic predicate must be "confusedly thought" in its subject. Peirce called this notion "monstrous," and claimed it could not be the case for mathematics:

> I can easily throw all the axioms of number, which are neither numerous nor complicated, into the antecedent of a proposition – or into its subject if that be insisted upon – so that the question of whether every number is the sum of three cubes, is simply a question of whether that is involved in the conception of the subject and nothing more. But to say that because the answer is involved in the conception of the subject, it is confusedly thought in it, is a great error. (4.86)

Now Peirce did not accept Kant's view that mathematics is synthetic, at least not in the way Kant intended it (3.560, 4.91, 4.232). But even analytic predicates, according to Peirce, can be "as utterly hidden as gold ten feet below ground." Thus the mathematician cannot obtain his conclusions by any "strain of mental vision." He must be willing to actively engage his material, to manipulate formulae and diagrams, to experiment on them and observe their behavior. It is in this sense that Peirce considered mathematics experiential. The question of how analytic predicates are obtained, however, is different from the question of why such predicates are necessary. Peirce criticized Kant for assimilating the former question to the latter. In a like fashion, Goudge apparently assimilates the latter to the former – or else he would not interpret Peirce's remarks as bearing upon the *justification* of mathematical assertions. Mathematics, for Peirce, is not experiential in the sense of being constructivist, of being inductive rather than deductive, or of being synthetic rather than analytic. The passages in which Peirce spoke of the importance of observation and experiment dealt only with the *discovery* of mathematical theorems. As Murphey puts it, "the

---

[88]We are in agreement with Murphey's discussion of this issue. See [**70**, pp. 232–235].

[89]Goudge's neglect of this distinction (at 4.85) is especially ironic in view of the fact that he later says: "It seems a pity that Peirce did not 'prescind' more carefully the logical from the psychological features of mathematical inquiry." [**38**, p. 73].

emphasis on observation is necessary to explain why our deductions surprise us, not why they are deductions."[90]

Peirce's overall conception of mathematics was *not* intuitionist. He was actually on the opposite end of the philosophical spectrum from intuitionism in terms of his enthusiasm for Cantor and the theory of transfinite numbers. Nevertheless, there was one particular respect in which Peirce's philosophy did resemble intuitionism: in its view of the relation between mathematics and logic.

Although Peirce would not have accepted Brouwer's description of mathematics as "essentially languageless," he would have agreed with Brouwer on the independence of mathematics from logic. "The recognition of mathematical necessity," he said. "is performed in a perfectly satisfactory manner antecedent to any study of logic." (2.191) Peirce was quite definite on this point. Whenever mathematics seems to appeal to logic, it is really appealing to that part of logic "which consists merely in an application of mathematics, so that the appeal will be, not of mathematics to a prior science of logic, but of mathematics to mathematics." (1.247) Even when there is a dispute, over whether some conclusion does or does not follow from given premises, "an appeal in mathematics to logic could only embroil a situation." (4.243)

The independence of mathematics – the science *which draws* necessary conclusions – from logic – the science *of drawing* necessary conclusions – Peirce sometimes expressed in terms of theory and practice:

> ...just as it is not necessary, in order to talk, to understand the theory of the formation of vowel sounds, so it is not necessary, in order to reason, to be in possession of the theory of reasoning.[91]

But Peirce was not really comfortable with the application to reasoning of this distinction. Any reasoning, insofar as it can be called good or bad, must involve some measure of deliberate control:

> Any operation which cannot be controlled, any conclusion which is not abandoned, not merely as soon as *criticism* has pronounced against it, but in the very act of pronouncing that decree, is not of the nature of rational inference – is not reasoning. (5.108)

---

[90]See [**70**, p. 131].
[91]4.242. Cf. [**70**, pp. 229–230].

To reason is to approve of one's reasoning. And such approval, Peirce said, "cannot be deliberate unless it is based upon the comparison of the thing approved with some idea of how such a thing ought to appear." (2.186) Thus all reasoning requires the possession of *some* theory of reasoning. (See 5.108, 5.130) Consequently, Peirce preferred to speak of the distinction, not between theory and practice, but between *logica docens* and *logica utens*. *Logica docens* is basically what Peirce called "Critic," the classification of arguments "as the result of scientific study." (2.204) *Logica utens* is likewise a theory as to the proper classification of arguments – but such a theory, Peirce said, as is gotten through instinct and experience "antecedent to any systematic study of the subject." (2.204) So the relation between *logica docens* and *logica utens* is roughly like that between a science and the uncodified rules by which an art is practiced.[92]

Mathematics, according to Peirce, operates entirely on the basis of the latter:

> ... mathematics performs its reasonings by a *logica utens* which it develops for itself, and has no need of any appeal to *logica docens* for no disputes about reasoning arise in mathematics which need to be submitted to the principles of the philosophy of thought for decision. (1.417)

Although less systematic than *logica docens*, the *logica utens* of mathematics is the only court of appeal for mathematical disputes. It would be fruitless to refer such disputes to *logica docens* because "true mathematical reasoning is so much more evident than it is possible to render any doctrine of logic proper." (4.243) It is this quality of 'evidence' or 'transparency' that renders mathematics independent of logic. Peirce sometimes explained it by the fact that mathematics deals solely with creations of the mind:

> Nor is the reason for this immunity of mathematics far to seek. It arises from the fact that the objects which the mathematician observes and to which his conclusions relate are objects of his mind's own creation. (3.426) Mathematical reasoning holds. Why should it not? It elates only to the creations of the mind, concerning which there is no obstacle to our learning whatever is true of them. (2.192)

---

[92]A description of *logica docens* and *logica utens* as, respectively, the "science" and "art" of logic can be found in DeMorgan [**21**, p. 84].

Now mathematical objects are created by the mind, not in the sense of being constructed out of intuitions, but in the sense of being suppositions entertained by the mind without reference to what might or might not actually be the case. It is clear from the following passage that this is the interpretation Peirce intended:

> Mathematical reasoning derives no warrant from logic. It needs no warrant. It is evident in itself. It does not relate to any matter of fact, but merely to whether one supposition excludes another. Since we ourselves create these suppositions, we are competent to answer them. But it is when we pass out of the realm of pure hypothesis into that of hard fact that logic is called for. (2.191)

So the quality of 'evidence' in mathematical reasoning, and hence its immunity to *logica docens*, is the result of its being purely hypothetical, untainted by factual considerations:

> Mathematics is engaged solely in tracing out the consequences of hypotheses. As such, she never at all considers whether or not anything be existentially true.... But now suppose that mathematics strikes upon a snag; and that one mathematician says that it is evident that a consequence follows from a hypothesis, while another mathematician says it evidently does not. Here, then, the mathematicians find themselves suddenly abutting against brute fact; for certainly a dispute is not a rational consequence of anything. True, this fact, this dispute, is no part of mathematics...However, because this dispute relates merely to the consequence of a hypothesis, the mere careful study of the hypothesis, which is pure mathematics, resolves it; and after all, it turns out that there was no occasion for the intervention of a science of reasoning. (1.247)

In particular, the hypothetical nature of mathematical reasoning is responsible for some of its other important characteristics, such as, e.g., "the peculiar difficulty, complication, and stress of its reasonings; the perfect exactitude of its results;..." (4.237) In a similar vein, Peirce tied the intricacy of mathematical reasoning to the "familiarity" of its concepts:

> ...in mathematics, the reasoning is frightfully intricate, while the elementary conceptions are of the last degree of familiarity; in contrast to philosophy [including logic] where the reasonings

are as simple as they can be. while the elementary conceptions
are abstruse and hard to get clearly apprehended. (3.560)

Unencumbered by the factual demands placed upon logic and philosophy, mathematicians become especially adept at sheer reasoning. Peirce called them, "the
very best reasoners in the world." (Ms 316a–s) Thus it is no wonder that Peirce
considered mathematical reasoning to fall outside the jurisdiction of *logica docens*. Peirce actually maintained the opposite viewpoint, viz., that "such difficulties as may arise concerning necessary reasoning have to be solved by the
logician by reducing them to questions of mathematics."[93]

Peirce not only held mathematics to be independent of logic in principle,
he claimed that this independence was, in fact, confirmed by history:

> ...no case is adducible in which the science of logic has availed
> to set mathematicians right or to save them from tripping. On
> the contrary, attention once having been called to a supposed
> inferential blunder in mathematics, short time has ever elapsed
> before the whole mathematical world has been in accord, either
> that the step was correct, or else that it was fallacious; and this
> without appeal to logic, but merely by the careful review of the
> mathematics as such. Thus, historically mathematics does not,
> as a priori it cannot, stand in need of any separate science of
> reasoning. (1.248)

Elsewhere Peirce admitted that a few prolonged disputes have indeed occurred
in mathematics, but only as a consequence of the importation of extra-
mathematical considerations. "Never, in the whole history of the science," he
said, "has a question whether a given conclusion followed *mathematically* from
given premises, when once started, failed to receive a speedy and unanimous
reply."[94]

Peirce did not mean by this to suggest that mathematical reasoning is
infallible. On the contrary, "It is fallible. as everything human is fallible."
(2.193) As anyone who has ever added a column of figures can testify, it is not
difficult to make a mathematical blunder. But such blunders are assented to
only insofar as they remain undetected. "In mathematics errors of reasoning
have occurred, nay, have passed unchallenged for thousands of years. This,
however, was simply because they escaped notice." (4.243) So Peirce described
mathematics as having a sort of "practical infallibility." (1.248) "There is no

---

[93]4.243. Also see 3.427.
[94]4.243. Also see 3.426.

more satisfactory way of assuring ourselves of anything," he said, "than the mathematical way of assuring ourselves of mathematical theorems." (2.193)

We have been concerned, so far, with the question of whether mathematics is independent of logic. The intuitionists actually took the further step of asserting that logic depends upon mathematics. Heyting, for example, said that logic is merely a *description* of certain general features of mathematical inference. Insofar as he thought that the science of drawing necessary conclusions depended upon the science which draws them, Peirce also took this step. "Logic can be of no avail to mathematics," he said, "but mathematics lays the foundation on which logic builds," (2.197) and, "Mathematics is not subject to logic. Logic depends upon mathematics." (2.191) Moreover, Peirce did not merely construe this dependence as "existential."[95] He conveyed the impression that mathematical reasoning is the *ideal* or *form* upon which all necessary reasoning is based:

> Deduction is the only necessary reasoning. It is the reasoning
> of mathematics. ...I declare that all necessary reasoning, be
> it the merest verbiage of the theologians, so far as there is any
> semblance of necessity in it, is mathematical reasoning. (5.145,
> 5.148)

In this respect, logic depends upon mathematics not only for its existence, but also for its model of deductive inference. This is what Peirce meant by suggesting that "logic ought to draw upon mathematics for control of disputed principles." (3.427) In one passage, Peirce even said of mathematics that "the only concern that logic has with this sort of reasoning is to *describe* it."[96]

Now it might be objected that this conflicts with Peirce's contention that logic is a normative science. But it is clear that Peirce did not consider logic normative in the sense of being qualified to evaluate, or prescribe to, mathematics. Furthermore, Peirce's similarity to the intuitionists is restricted to the issue of how mathematics is *related* to logic. It should not be extended to their overall conceptions of logic, any more than to their overall conceptions of mathematics. Although Peirce would have agreed with Heyting that logic does describe mathematics, he would not have agreed that this is all, or even most, of what logic does; he would have balked at the supposition that logic *merely* describes mathematics. Peirce's understanding of logic was broader than that of the intuitionists. As well as "Critic," it included "Speculative Grammar" and "Speculative Rhetoric," and within "Critic," not only deduction, but also

---

[95]See [**38**, pp 57, 58].
[96]2.192, emphasis added.

induction and abduction. (4.8, 2.619–2.623) But even if we confine ourselves to the narrowest meaning of the term "logic" – the study of deductive inference – we would, according to Peirce, be dealing with a science which is categorical instead of hypothetical.[97] Logic is primarily concerned, not with the necessity of its conclusions, but with their *truth*:

> ... the hypotheses from which the deductions of normative science proceed are *intended to conform* to positive truth of fact and those deductions derive their interest from that circumstance almost exclusively; while the hypotheses of pure mathematics are purely ideal in intention, and their interest is purely intellectual. (5.126)

For Peirce, all logic is oriented toward experience, toward phenomena. But like ethics and aesthetics, and in contrast to phenomenology, logic treats of the conformity of phenomena to certain *ends* (in the case of logic, to the end of "representing something." See 5.123). So logic is normative by virtue of its connection, not to mathematics, but to what is neutrally given:

> ... there is a most intimate and essential element of Normative Science which is still more proper to it, and that is its *peculiar appreciations*, to which nothing at all in the phenomena, in themselves, corresponds. These appreciations relate to the conformity of phenomena *to ends* which are not immanent within those phenomena. (5.126)

That it has these peculiar appreciations does serve to distinguish logic from mathematics. But the appreciations themselves are of phenomena. There is no reason, then, why logic should not be both a normative science and purely descriptive with regard to mathematical reasoning.[98]

---

[97]Although Peirce only distinguished two senses in which he used the term "logic," viz., as "general semeiotic" and as "Critic" (1.444), he clearly employed the term in this third, even narrower, sense as well.

[98]On logic as a normative science, see 1.573–1.584, 5.108–5.150, and [**89**, pp. 25–51].

Peirce's classification of the sciences will shed some further light on his understanding of the relation between mathematics and logic. Peirce divided the sciences into theoretical and practical branches; and within the former he further divided them into sciences of discovery and sciences of review. The sciences of discovery were classified as follows:[99]

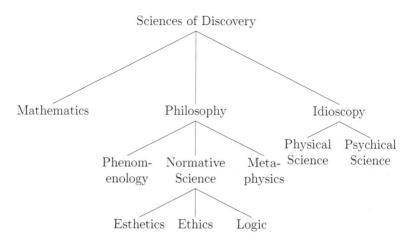

Peirce ordered the different sub-divisions in accordance with his categories, so that the entire scheme is hierarchical, "a sort of ladder descending into the well of truth."[100] Each science depends upon those to the left it, and is independent of those to the right of it. For instance: Peirce remarked that there are five disciplines which "do not more or less depend upon the science of logic." (2.120) These are mathematics, phenomenology, aesthetics, ethics, and logic itself. All the others, he said, "draw principles from the theory of logic." (2.120–21) Mathematics, however, stands at the vanguard of the sciences. Thus Peirce said it applied to every other science "without exception," but that "this is not true of any other science, since pure mathematics has not, as a part of it, any application of any other science." (1.245) In this sense, Peirce's conception of the relation between mathematics and logic was a particular instance of his wider position on the self-sufficiency of mathematics.[101]

---

[99]1.239–43. This is taken from Peirce's 1902 classification. His earlier 1889 classification also placed mathematics ahead of logic. See 3.427.

[100]2.119. On Peirce's use of the categories in Classification, see [**89**, pp. 18–24].

[101]This self-sufficiency resembles the description by Benjamin Peirce who wrote, at the beginning of *Linear Associative Algebra*, that "mathematics, according to this definition, belongs to every enquiry, physical as well as moral" and explained, "the branches of mathematics are as various as the sciences to which they belong, and each subject of physical enquiry has its appropriate mathematics. In every form of material manifestation, there is a corresponding form of human thought, so that the human mind is as wide in its range of

3.3.4. Formalism. Essentially, formalism viewed mathematics as the science of formal axiomatic-deductive systems.[102] We are in agreement with Murphey that Peirce's general orientation toward mathematics was more formalist than anything else.[103] This is implied, for example, by Peirce's treatment of mathematics as both "purely hypothetical" and deductive:

> Mathematics is the study of what is true of hypothetical states
> of things. That is its essence and definition. Everything in
> it. therefore. beyond the first precepts for the construction of
> the hypotheses, has to be of the nature of apodictic inference.
> (4.233)

Formalism is also suggested by Peirce's axiomatization programs of 1867 and 1881, especially by his reference, in both papers, to his axioms as "definitions."[104] What Peirce meant by this was that only the formal relationships given in the axioms really define the primitive "objects," e.g., numbers, being considered by the mathematician:

> ...In pure algebra, the symbols have no other meaning than
> that which the formulae impose upon them. In other words,
> they signify any relations which follow the same laws. Any-
> thing more definite detracts needlessly and injuriously from
> the generality and utility of the algebra. (4.314)
> A proposition is not a statement of perfectly pure mathematics
> until it is devoid of all definite meaning, and comes to this –
> that a property of a certain icon is pointed out and is declared
> to belong to anything like it, of which instances are given.
> (5.567)

Like the formalists, Peirce considered the axioms and theorems of a mathematical system to provide "implicit definitions" of the primitive notions.

There was also a different sense in which Peirce thought of axioms as "definitions," viz., in that they "define" the branch of mathematics being formalized:

---

thought as the physical universe in which it thinks. The two are wonderfully matched." [**76**, p. 98].

[102]There was much more to formalism than this, of course, but an examination of the position as elaborated by Hilbert and Curry lies outside the scope of this section. See [**56**, pp. 72–97].

[103]See [**70**, pp. 235–237].

[104]Goudge is mistaken in regarding these references as an indication of Peirce's logicism. See [**38**, p. 57] and above, pp. 118ff.

It is quite true, . . . that a postulate, after all, is merely a part of the definition of the underlying conception of the branch of mathematics to which it refers, (Euclid's celebrated fifth postulate, for example, being merely a part of the definition of Euclidean space) . . . (4.325)

According to Bernays, it was only in this latter sense that Hilbert spoke of axioms as definitional:

The axioms generally impose conditions on the relations and on the kinds of elements of the system; of the relations or the kinds of elements, others characterize the space with respect to the elements and relations. The entire axiom system – as Hilbert observed in a letter to Gottlob Frege – can be regarded as a single definition. But this is an *explicit* definition of a term denoting the relational structure in question.[105]

It is evident, though, that the two senses are related; that it is by characterizing the expected behavior of certain primitives, for instance, that an axiom system is accepted as having captured a traditional branch of mathematics. Curry, in fact, explicitly connects them:

The primitive frame specifies . . . which elementary propositions are true, and therefore determines the meaning of the fundamental predicates, In this sense the primitive frame defines the system.[106]

In both cases, Peirce meant to underline the fact that mathematical concepts should be explained in terms of the formal system associated with them, instead of by extra-systematic considerations such as experience and intuition.

Despite our larger agreement, we have some reservations about Murphey's criticism of Peirce for having neglected the question of consistency.[107] The formalist emphasis upon consistency proofs was another attempt to secure arithmetic (and analysis) – by attaching it, from the outside as it were, to constructive reasoning. But Peirce did not think that arithmetic needed any such securing. "It will be time to defend it," he said, "when it shall once be deliberately doubted." (2.197) On the other hand, Peirce would surely have accepted formalist metamathematics – remaining suspicious of its original motivation –

---

[105]Paul Bernays, s.v. "Hilbert, David."
[106]See [19].
[107]See [70, pp. 236–237].

as a legitimate arena for mathematical inquiry. We do not think, for example, that his objection to "what people call an 'interpretation"' would extend to the study of formal models.[108]

One might well wonder how Peirce could have been a formalist and a realist at the same time. The answer is rather simple. Peirce's formalism had to do with his understanding of the proper business of mathematics. His realism, however, was a metaphysical doctrine, formulated in response to concerns that lay outside the province of mathematics. As a philosopher, for instance, Peirce would undoubtedly have maintained that there *are* numbers such as described by his 1881 axiom system.[109] But he would not have considered this metaphysical position relevant to the mathematician:

> The pure mathematician deals exclusively with hypotheses. Whether or not there is any corresponding real thing, he does not care. His hypotheses are creatures of his own imagination; but he discovers in them relations which surprise him sometimes. A metaphysician may hold that this very forcing upon the mathematician's acceptance of propositions for which he was not prepared, proves, or even constitutes, a mode of being independent of the mathematician's thought, and so a reality. But whether there is any reality or not, the truth of the pure mathematical proposition is constituted by the impossibility of ever finding a case in which it fails. (5.567)

If generally adopted, Peirce's severance of mathematics from metaphysics, even correct metaphysics, would have helped eliminate the methodological disputes that dominated foundations at the turn of the century. In this respect, he seems to have anticipated the more pragmatic technical approach of recent years.

Peirce's view of the relation between mathematics and logic was similar to the intuitionists, and his overall view of mathematics was probably closer to formalism, but he differed from logicism, intuitionism, and formalism alike in his conception of the total independence and integrity of mathematical inquiry. For Peirce, mathematics is a First. It requires no epistemological foundation, but is itself the foundation for other knowledge.

---

[108]4.130. See [**70**, p. 237] and the discussion of Peirce on the axiom of infinity, above, pp. 69, 70.

[109]See 4.161, where Peirce describes numbers platonically as "eternal beings" of the "Inner World."

CHAPTER 4

Letters

This appendix contains several letters from Max Fisch which should be of interest to Peirce scholars. The first letter, dated July 14, 1976, is addressed to John Smith. As I remember, John, with whom I was studying at the time, offered to send Max Fisch some of my questions – and John must have just given me the return letter. The seed for this book is clearly contained in the first paragraph of this letter – in Max's reference to Gerrit Mannoury. The next two letters, dated in 1981, are addressed to me and give Max's initial response to my dissertation.

4.1. Max Fisch to John Smith, July 14, 1976

INDIANA
UNIVERSITY
PURDUE
UNIVERSITY   at INDIANAPOLIS

PEIRCE EDITION PROJECT
University Libraries • 420 Blake Street • 46202

July 14, 1976

Dear John:

I have your note of the 11th. With regard to the passage in CP 4.331,
I think that, as Hartshorne and Weiss assume in note ||, the reference is
to Schröder's article. Gerrit Mannoury followed Schröder in assigning
priority to Peirce ("De zogenaamde grondeigenschap der rekenkunde," Handelingen
van het 8ste Nederl. Natuur- en Geneeskundig Congres, pp. 121-147, 1901;
German translation: Methodologisches und Philosophisches zur Elementar-
mathematik, Haarlem: P. Visser Azn., 1909, see esp. 75-77 n. 3; Peirce
received a copy of the Dutch original, almost certainly from Mannoury himself).

I assume that Schröder's article was among the four he sent Peirce
early in December 1898 (see inclosed sheet).

Hartshorne and Weiss are mistaken in their first note. CP 4.331-344 are
from a paper read to the National Academy of Sciences at its meeting held
at Columbia University, November 15-16, 1904 (Ms 95 in Robin's Catalogue).
4.345-340 are from Ms 44. (Robin's account of Ms 95 is misleading.)

We are grateful for your support of our Project!

Sincerely,

Max

Max H. Fisch

Medicine • Dentistry • Nursing • University Hospitals • Law • Social Service • Liberal Arts
Engineering and Technology • Fine Arts • Business • Education • Science • Physical Education

4.2. Max Fisch to Author, August 21, 1981

INDIANA UNIVERSITY-
PURDUE UNIVERSITY
at INDIANAPOLIS

PEIRCE EDITION PROJECT
University Library
420 Blake Street
Indianapolis, Indiana 46202

August 21, 1981

Dear Paul:

I have been reading some parts of your thesis more closely, with
admiration and profit.

If you publish anything involving mention of Benjamin Peirce's
Linear Associative Algebra, it might be well to mention the original
lithographed edition (Washington, D. C., 1870), which Archibald describes
as "developed from papers read before the National Academy of Sciences,
1866-1870."

Charles Peirce in CP 4.301n1 attaches importance to the fact that
he was working on the logic of relations at the same time. Cf. CP 4.239
(top of p. 199), 533, 614. Since you make something of these passages,
it may be important to some of your readers to associate them with the
years immediately preceding 1870 rather than those shortly preceding
1881, especially as his father died in 1880 and they had been seeing
much less of each other since 1875.

BP was already working on linear associative algebra at the time
of CSP's paper of 1867 "Upon the Logic of Mathematics."

There have been further delays, and our volume 1 will not appear
before late fall. And it is still the case, as I wrote you on April 27,
that "Most of my work on the introduction to vol. 2 (1867-1871) is still
to be done, so if you have further thoughts about 'Upon the Logic of
Mathematics' they may still make a contribution to the volume. (In any
case, we'll count on you for 'On the Logic of Number' etc.)"

Any further developments careerwise?

All the best to you!

Max

Max H. Fisch

I've had no word from Donna since I saw her so briefly in December.

4.3. Max Fisch to Author, August 22, 1981

The Charles S. Peirce Society

August 22, 1981

Dear Paul:

Writing at home and having no Project stationery here, but some CSP
Society stationery, I use the latter.

I notice that you do mention the 1870 edition of <u>Linear Associative
Algebra</u> on p. 156. Another relevant passage is in a late letter to James
published by Eisele in <u>New Elements</u> 3:836-866 at 855.

Some notes on your p. 139n1: Peirce called on De Morgan in London in July
1870. Papers he sent to De Morgan bear De Morgan's notes of dates of receipt.
Peirce met Jevons about the same time in 1870. A letter from Peirce to Jevons
dated Peath 1870 Aug 25 is among the Jevons Papers at Manchester. It will be
published in our second volume. Carolyn Eisele was quoting from my "Chronicle
of Pragmaticism" (<u>Monist</u> 48: 441-466, 1964, at 461 under October 29).

<u>Linear Associative Algebra</u>'s first edition is treated as a published book
in most catalogues. The AJM edition of 1881 was also published as a separate
volume by Van Nostrand in 1882. My own copy is of the Van Nostrand edition.

Your p. 140 brings home to me the embarrassing fact that I have not yet
written out such accounts of the 1870 and 1881 editions as I must for the
CSP biography. The National Academy of Sciences intended to publish the 1870
edition in its Memoirs but lacked the necessary funds at the time, and Hilgard
of the Coast Survey produced the lithographed edition. I assume the necessary
information concerning the 1881 edition is in my files but it would take some
time to dig it out. As of today, I don't remember whether BP himself offered
it to Sylvester for the AJM before his death in 1880, or whether CSP after BP's
death arranged with Sylvester for a posthumous edition edited by himself. Nor
do I remember when CSP offered his own paper to Sylvester, or how the decision
was reached to publish BP after CSP. I must work up my account of the 1870
edition for the introduction to our vol. 2, and of the 1881 edition for the
introduction to vol. 4.

I admire more and more the job you have done and hope you will get some
of it into print soon.

Sincerely yours,

Max I. Fisch

Bibliography

[1] Alexander Abian, The Theory of Sets and Transfinite Arithmetic, W. B. Saunders Co., Philadelphia, 1965.

[2] Stephen F. Barker, Philosophy of Mathematics, Prentice-Hall, Englewood Cliffs, 1964.

[3] E. T. Bell, The Development of Mathematics, McGraw-Hill, New York, 1940.

[4] Paul Benacerraf and Hilary Putnam (eds.), Philosophy of Mathematics, Prentice-Hall, Englewood Cliffs, 1954.

[5] George Allen Benedict, The Concept of Continuity in Charles Peirce's Synechism, Ph.D. thesis, State University of New York at Buffalo, 1973.

[6] George Berkeley, An Essay Towards a New Theory of Vision, Tonsan, Dublin, 1709.

[7] _____, The Analyst, or A Discourse Addressed to an Infidel Mathematician, Tonsan, London, 1734, excerpted: [**72**, p. 286–295].

[8] George Berkeley and Colin Turbayne, A Treatise Concerning the Principles of Human Knowledge, Bobbs-Merrill, New York, 1965.

[9] Paul Bernays, Hilbert, David, Encyclopedia of Philosophy (Paul Edwards, ed.), Macmillan, New York, 1972, reprint ed. A. N. Prior, pp. 496–504.

[10] Evert W. Beth, The Foundations of Mathematics, North-Holland, Amsterdam, 1959.

[11] _____, Mathematical Thought, D. Riedel, Dordrecht, The Netherlands, 1965.

[12] Bernard Bolzano, Paradoxien des Unendlichen, Fr. Prihonsky, Leipzig, 1851, translation by Donald A. Steele as Paradoxes of the Infinite, London:Routledge & Kegan Paul, 1950.

[13] George Boolos, To Be is to be a value of a variable (or to be some values of some variables), Journal of Philosophy **81** (1984), 430–449.

[14] Carl Boyer, The History of the Calculus and its Conceptual Development, Dover, New York, 1959.

[15] L. E. J. Brouwer, Historical background, principles and methods of intuitionism, South African J. Sci. **49** (1952), 139–146.

[16] Cesare Burali-Forti, Una questione sui numeri transfiniti, Rendiconti di Palermo **11** (1897), 154–164, translated: "A Question on Transfinite Numbers," [**43**, pp. 104–111].

[17] Georg Cantor, Über die Ausdehnung eines Satzes aus der Theorie der trigonometrischen Reihen, Math. Ann. **5** (1872), no. 1, 123–132.

[18] _____, Beiträge zur Begründung der transfiniten Mengenlehre, Math. Ann. **49** (1897), no. 2, 207–246, translated with an Introduction and Notes by Philip E. B. Jourdain: Contributions to the Founding of the Theory of Transfinite Numbers. Open Court Publishing Co., 1915; New York:Dover, 1955.

[19] Haskell B. Curry, Remarks on the definition and nature of mathematics, Dialectia **48** (1954), 228–233, reprinted: [**4**, pp. 152–56].

[20] Tobias Dantzig, Number: The Language of Science, Doubleday, Garden City, New York, 1954.

[21] Augustus De Morgan, On the Syllogism and other Logical Writings, Routledge & Kegan Paul, 1966, edited by Peter Heath.

[22] Shannon Dea, 'Merely a veil over the living thought': Mathematics and Logic in Peirce's
      Forgotten Spinoza Review, Transactions of the Charles S. Peirce Society **42** (2006),
      no. 4, 501–517.
[23] Richard Dedekind, Stetigkeit sind irrationale Zahlen, Braunschweig, Vieweg, 1872,
      translated as Continuity and Irrational Numbers by W. W. Beman in Essays on Number
      Theory, Open Court, Chicago, 1901, reprint: New York:Dover, 1963, pp. 1–27.
[24] _____, Was sind und was sollen die Zahlen?, Braunschweig, Vieweg, 1888, translated as
      The Nature and Meaning of Numbers, by W. W. Beman in Essays on Number Theory,
      Open Court, Chicago, 1901, reprint: New York:Dover, 1963, pp. 31–115.
[25] Paul du Bois-Reymond, Versuch einer Classification der Willkürlichen Functionen
      Reeller Argumente nach ihren Aenderungen in den Kleinsten Intervallen, J. Reine
      Angew. Math. **79** (1875), 21–37.
[26] Dummett, Wittgenstein's philosophy of mathematics, The Philosophical Review **68**
      (1959), 324–348.
[27] Carolyn Eisele, The Charles S. Peirce – Simon Newcomb Correspondence, Proceedings
      of the American Philosophical Society, vol. 101 (5), M'Calla & Stavely, Philadelphia,
      31 October 1957, pp. 409–433.
[28] _____, Fermatian Inference and De Morgan's Syllogism of Transposed Quantity in
      Peirce's Logic of Science, Physis, Rivista di storia della scienza **5** (1963), no. 2, 120–
      128.
[29] _____, The new elements of mathematics by Charles S. Peirce, Men and Institutions
      in American Mathematics, no. 13, Graduate Studies, Texas Tech University, College
      Station, 1976, pp. 111–121.
[30] Max Fisch and Atwell Turquette, Peirce's triadic logic, Trans. Charles S. Peirce Soc. **2**
      (1966), 71–85.
[31] Abraham Fraenkel and Yohoshua Bar-Hillel, Foundations of Set Theory, North-Holland,
      Amsterdam, 1958.
[32] Gottlob Frege, Begriffsschrift, eine der arithmetischen nachgebildete Formelsprache des
      reinen Denkens, L. Nebert, Halle, 1879, translated by Stephen Bauer-Mengelberg: [**43**,
      pp. 1–82].
[33] _____, Die Grundlagen der Arithmetik, Jena, Breslau, 1884, translation by J. L.
      Austin as The Foundations of Arithmetic, New York:New York Philosophical Library,
      1950.
[34] _____, Grundgesetze der Arithmetik, Hermann Pohle, Jena, Vol. 1, 1893; Vol. 2 1903,
      translation by P. E. B. Jourdain, J. Stachelroth, Max Black, and P. T. Geach in Philo-
      sophical Writings of Gottlob Frege, ed. by Geach and Black, Oxford:Basil Blackwell.
      1970.
[35] Benjamin Ives Gilman, On the properties of a one-dimensional manifold, Mind **1** (1892),
      518–526.
[36] Kurt Gödel, Einige metamathematische Resultate über Entscheidungsdefinitheit und
      Widerspruchsfreiheit, Anzeiger der Akademie der Wissenschaften in Wein **67** (1930),
      214–215, translated as "Some Metamathematical Results on Completeness and Consis-
      tency": [**43**, pp. 595–596].
[37] _____, Über formal unentscheidbare Sätze der Principia Mathematica und verwandter
      Systeme. I, Monatsh. Math. **38** (1931), 173–198, translated as "On Formally Undecid-
      able Propositions of Principia Mathematica and Related Systems I": [**43**, pp. 596–616].
[38] Thomas A. Goudge, The Thought of C. S. Peirce, University of Toronto Press, Toronto,
      1950, reprint ed., New York:Dover, 1969.
[39] Hermann Grassmann, Lehrbuch der Arithmetik für höhere Lehranstalten, Th. Chr. Fr.
      Enslin, Berlin, 1861.
[40] Ivor Grattan-Guinness, Dear Russell-Dear Jourdain, Gerald Duckworth, London, 1977.

[41] Paul R. Halmos, Naive Set Theory, Springer-Verlag, New York, 1974, Reprint of the 1960 edition, Undergraduate Texts in Mathematics.

[42] Jérôme Havenel, Peirce's Clarifications of Continuity, Transactions of the Charles S. Peirce Society **44** (2008), no. 1, 86–133.

[43] Jean van Heijenoort (ed.), From Frege to Gödel: A Source Book in Mathematical Logic, 1879–1931, Harvard University Press, Cambridge, 1967.

[44] ———, Gödel's Theorem, Encyclopedia of Philosophy (Paul Edwards, ed.), vol. 3, Macmillan, New York, 1972, reprint ed. Paul Bernays, pp. 348–357.

[45] Bertrand Helm, The Critical Philosophy and the Royce-Bradley Dialogue, Journal of the History of Philosophy **11** (1973), 229–236.

[46] C. G. Hempel, On the Nature of Mathematical Truth, Amer. Math. Monthly **52** (1945), 543–556, reprinted: [**4**, 366–381].

[47] Leon Henkin, Completeness in the Theory of Types, J. Symbolic Logic **15** (1950), 81–91, reprinted: [**51**, pp. 51–63].

[48] Arend Heyting, Intuitionism. An Introduction, North-Holland, Amsterdam, 1956.

[49] David Hilbert, Über die Grundlagen der Logik und der Arithmetik, Taubner, Leipzig, 1905, translation by Beverly Woodward: [**43**, pp. 129–138].

[50] ———, Über das Unendliche, Math. Ann. **95** (1926), no. 1, 161–190, partial translation as "On the Infinite": [**4**, 134–151].

[51] Jaakko Hintikka (ed.), The Philosophy of Mathematics, Oxford University Press, 1969.

[52] E. V. Huntington, The Continuum and Other Types of Serial Order, Harvard University Press, Cambridge, 1917.

[53] Hubert Kennedy (ed.), Selected Works of Giuseppe Peano, University of Toronto Press, Toronto, 1973.

[54] R. B. Kershner and L. R. Wilcox, The Anatomy of Mathematics, The Ronald Press, New York, 1950.

[55] Morris Kline, Mathematical Thought from Ancient to Modern Times, Oxford University Press, New York, 1972.

[56] Stephan Körner, The Philosophy of Mathematics, Harper & Row, New York, 1962.

[57] Bruce Kuklick, Rise of American Philosophy, Yale University Press, New Haven, 1977.

[58] Casimir Kuratowski, Sur la notion de l'ordre dans la theorie des ensembles, Fundamenta Mathematicae **2** (1921), 161–171.

[59] C. I. Lewis and C. H. Langford, The Development of Symboic Logic, Contemporary Readings in Logical Theory (Irving Copi and James Gould, eds.), MacMillan, New York, 1967, pp. 3–20.

[60] David Lewis, Parts of Classes, Blackwell, London, 1991.

[61] A. H. Lightstone, Symbolic Logic and the Real Number System, Harper & Row, New York, 1965.

[62] A. Malcev, Untersuchungen aus dem Gebiete der mathernatischen Logik, Recueil Mathematigue **1** (1936), 323–336.

[63] Jerome H. Manheim, The Genesis of Point Set Topology, Pergamon Press, Oxford, 1964.

[64] Gerrit Mannoury, Methodologisches und Philosophisches zur Elementar-Mathematik, P. Visser Aznl, Haarlem, 1909.

[65] Elliott Mendelson, Introduction to Mathematical Logic, D. Van Nostrand, Princeton, 1964.

[66] Emily Michael, Peirce's Early Study of the Logic of Relations, 1865-1867, Trans. Charles S. Peirce Soc. **10** (1974), 63–75.

[67] John Stuart Mill, A System of Logic, London, 1843.

[68] E. Moore, Matthew, New Essays on Peirce's Mathematical Philosophy, Open Court, Chicago, 2010.

[69] Gregory Moore, Review of *From Frege to Gödel* by Jean van Heijenoort, Historia Mathematica **4** (1977), 468–471.

[70] Murray G. Murphey, Development of Peirce's Philosophy, Harvard University Press, Cambridge, 1961.

[71] Ernest Nagel and James Newman, Gödel's Proof, New York University Press, New York, 1958.

[72] James R. Newman, The World of Mathematics, Simon & Schuster, 1956.

[73] Fricke Ore and Emmy Noether (eds.), C. S. Peirce Gesammelte Mathematische Werke, vol. 1–3, Chelsea, Braunschweig, 1969, reprint ed.

[74] Henry A. Patin, Pragmatism, Intuitionism and Formalism, Philosophy of Science **4** (1957), 243–252.

[75] Giuseppe Peano, Arithmetices principia nova methodo exposita, Bocca, Turin, 1889, translated as "The Principles of Arithmetic, Presented by a New Method": [**53**].

[76] Benjamin Peirce, Linear Associative Algebra, Amer. J. Math. **4** (1881), no. 1-4, 97–229.

[77] Charles Sanders Peirce, Upon the Logic of Mathematics, Proceedings of the American Academy of Arts and Sciences **7** (1867), 402–412.

[78] _____, On the Algebra of Logic, Amer. J. Math. **3** (1880), 15–57.

[79] _____, Studies in Logic by Members of the Johns Hopkins University, Boston, 1883.

[80] _____, On the Algebra of Logic: A Contribution to the Philosophy of Notation, Amer. J. Math. **7** (1885), 180–202.

[81] _____, The Logic of Mathematics in Relation to Education, Educational Review **15** (1898), 209–216.

[82] _____, Nation **70** (1900), 267.

[83] _____, Nation **75** (1902), 94–96.

[84] _____, Collected Papers of Charles Sanders Peirce (Charles Hartshorne and Paul Weiss, eds.), vol. 1–6, Harvard University Press, Cambridge, 1931–1935.

[85] _____, Collected Papers of Charles Sanders Peirce (Arthur W. Burks, ed.), vol. 7–8, Harvard University Press, Cambridge, 1958.

[86] _____, The New Elements of Mathematics by Charles S. Peirce (Carolyn Eisele, ed.), Mouton and Humanities Press, The Hague and Atlantic Highlands, N.J., 1976, 4 Volumes.

[87] _____, Writings of Charles S. Peirce: A Chronological Edition (Peirce Edition Project, ed.), Indiana University Press, Bloomington and Indianapolis, 1982-2009, Volumes 1-8.

[88] Jules Henri Poincaré, Science and Method, Cambridge University Press, London, 1914, translated by Francis Maitland with a preface by Bertrand Russell.

[89] Vincent G Potter, Charles S. Peirce on Norms and Ideals, University of Massachusetts Press, Amherst, 1967.

[90] Vincent G. Potter, Normative Science and the Pragmatic Maxim, Journal of the History of Philosophy **5** (1967), no. 1, 41–53.

[91] Vincent G. Potter and Paul Shields, Peirce's Definitions of Continuity, Trans. Charles S. Peirce Soc. **13** (1977), 20–34.

[92] A. N. Prior, Logic, History of, Encyclopedia of Philosophy (Paul Edwards, ed.), vol. 4, Macmillan, New York, 1972, pp. 513–568.

[93] _____, Logic, Modal, Encyclopedia of Philosophy (Paul Edwards, ed.), vol. 5, Macmillan, New York, 1972, pp. 5–12.

[94] Hilary Putnam, Mathematics, Matter and Method. Philosophical Papers, Volume I, Cambridge University Press, Cambridge, 1972.

[95] Willard Van Orman Quine, Review of the *Collected Papers of Charles Sanders Peirce, Volume 3*, Isis **22** (1934), no. 1, 285–297.

[96] _____, New Foundations for Mathematical Logic, Amer. Math. Monthly **44** (1937), no. 2, 70–80.

[97] _____, Set Theory and its Logic, Harvard University Press, Cambridge, 1963.

[98] F. P. Ramsey, Foundations, Essays in Philosophy, Logic, Mathematics and Economics, Routledge & Kegan Paul, 1978, ed. D.H. Mellor.

[99] John Richardson, The Grammar of Justification, Sussex University Press, London, 1976.

[100] Bernard Riemann, Über die Hypothesen welche der Geometrie zugrunde liegen, Gesammelte Mathematische Werke, Teubner, Leipzig, 2 ed., 1892, Habilitationsschrift, Göttigen, 1854, pp. 272–287.

[101] Don Davis Roberts, The Existential Graphs and Natural Deduction, Studies in the Philosophy of Charles Sanders Peirce (Edward C. Moore and Richard S. Robin, eds.), University of Massachusetts Press, Amherst, 1964, pp. 109–121.

[102] Richard S. Robin, Annotated Catalogue of the Papers of Charles S. Peirce, University of Massachusetts Press, Worcester, 1967.

[103] _____, The Peirce Papers: A Supplementary Catalogue, Trans. Charles S. Peirce Soc. 7 (1971), 37–57.

[104] Abraham Robinson, Non-standard Analysis, North-Holland, Amsterdam, 1966.

[105] R. M. Robinson, The Theory of Classes. A Modification of von Neumann's System, J. of Symbolic Logic 2 (1937), 29–36.

[106] J. B. Rosser, Extensions of Some Theorems of Gödel and Church, J. of Symbolic Logic 1 (1936), 87–91.

[107] Josiah Royce, The World and the Individual, First, MacMillan, New York, 1899, reprint ed., New York:Dover 1959.

[108] Bertrand Russell, On the Notion of Order, Mind 10 (1901), 30–51.

[109] _____, Principles of Mathematics, Cambridge University Press, London, 1903, paperback ed., New York:W.W. Norton.

[110] _____, Mathematical logic as based on the theory of types, Amer. J. Math. 30 (1908), no. 3, 222–262, reprinted: [43, pp. 150–182].

[111] _____, Sur les axiomes de l'infini et du transfini, Comptes Rendus 2 (1911), 22–35, translated: [40, pp. 162–174].

[112] _____, Introduction to Mathematical Philosophy, second ed., George Allen & Unwin, London, 1919, 12th impression, 1967.

[113] Ernst Schröder, Über zwei Definitionen der Endlichkeit und G. Cantor'sche Sätze, Nova Acta, Abhandlung der Kaiserl. Leop.-Carol. Deutschen Akademie der Naturforscher 71 (1888), no. 6, 303–362.

[114] Henry Maurice Sheffer, A set of five independent postulates for Boolean algebras, with application to logical constants, Trans. Amer. Math. Soc. 14 (1913), no. 4, 481–488.

[115] Paul Shields, Peirce's Axiomatization of Arithmetic, Studies in the Logic of Charles S. Peirce (Nathan Houser, Don D. Roberts, and James van Evra, eds.), Indiana University Press, Bloomington, 1997, pp. 43–52.

[116] Thoralf Skolem, Über die Unmöglichkeit einer vollständigen Charakterisierung der Zahlen riehe mittels eines endlichen Axiomensystems, Norsk Matematisk Forenings Skrifter 2 (1933), no. 10, 73–82.

[117] Patrick Suppes, Axiomatic Set Theory, Dover, New York, 1972.

[118] Alfred Tarski, Sur les ensembles finis, Fundamenta Mathematicae 6 (1924), 45–95.

[119] Atwell Turquette, Peirce's Icons for Deductive Logic, Studies in the Philosophy of Charles Sanders Peirce (Edward C. Moore and Richard S. Robin, eds.), University of Massachusetts Press, Amherst, 1964, pp. 95–108.

[120] Jouko Väänänen, Second-order Logic and Foundations of Mathematics, The Bulletin of Symbolic Logic 7 (2001), no. 4, 504–520.

[121] Oswald Veblen, A system of axioms for geometry, Trans. Amer. Math. Soc. 5 (1904), no. 3, 343–384.

[122] John von Neumann, On the Introduction of Transfinite Numbers, reprinted: [**43**, pp. 347–354].

[123] Hao Wang, The Axiomatization of Arithmetic, J. of Symbolic Logic **22** (1957), 145–158.

[124] Hermann Weyl, Philosophy of Mathematics and Natural Science. Revised and Augmented English Edition Based on a Translation by Olaf Helmer, Atheneum, New York, 1963.

[125] Alfred N. Whitehead and Bertrand Russell, Principia Mathematica, Cambridge University Press, London, 1910, 2nd ed., 1927; paperback ed. to *56, 1962.

[126] Ludwig Wittgenstein, Philosophical Investigations, MacMillan, New York, 1953, translated by G. E. M. Anscombe and Elizabeth Anscombe.

[127] Ernst Zermelo, Untersuchungen über die Grundlagen der Mengenlehre I, Math. Ann. **65** (1908), 261–281, translated by Stefan Bauer-Mengelberg as "Investigations in the Foundations of Set Theory," [**43**, pp. 200–215].

Made in the USA
Lexington, KY
18 November 2012